# Cost Control in Foodservice Operations

*David K. Hayes*
*Jack D. Ninemeier*

| | |
|---|---|
| SENIOR EDITORIAL DIRECTOR | Justin Jeffryes |
| EXECUTIVE EDITOR | Todd Green |
| EDITORIAL ASSISTANT | Kelly Gomez |
| SENIOR MANAGING EDITOR | Judy Howarth |
| PRODUCTION EDITOR | Mahalakshmi Babu |
| COVER PHOTO CREDIT | © Peathegee Inc/Getty Images |

This book was set in 9.5/10.5 STIX Two Text by Straive™.
Published by John Wiley & Sons, Inc., Hoboken, New Jersey.
Published simultaneously in Canada.
This book is printed on acid-free paper.

Founded in 1807, John Wiley & Sons, Inc. has been a valued source of knowledge and understanding for more than 200 years, helping people around the world meet their needs and fulfill their aspirations. Our company is built on a foundation of principles that include responsibility to the communities we serve and where we live and work. In 2008, we launched a Corporate Citizenship Initiative, a global effort to address the environmental, social, economic, and ethical challenges we face in our business. Among the issues we are addressing are carbon impact, paper specifications and procurement, ethical conduct within our business and among our vendors, and community and charitable support. For more information, please visit our website: www.wiley.com/go/citizenship.

ISBN: 978-1-394-20804-3 (PBK)

*Library of Congress Cataloging-in-Publication Data*

Names: Hayes, David K., author. | Ninemeier, Jack D., author. | John Wiley
& Sons, publisher.
Title: Cost control in foodservice operations / David K. Hayes, Jack D.
Ninemeier.
Description: Hoboken, New Jersey : Wiley, [2024] | Includes index.
Identifiers: LCCN 2023041646 (print) | LCCN 2023041647 (ebook) | ISBN
9781394208043 (paperback) | ISBN 9781394208081 (adobe pdf) | ISBN
9781394208098 (epub)
Subjects: LCSH: Food service—Cost control.
Classification: LCC TX911.3.C65 H38 2024 (print) | LCC TX911.3.C65
(ebook) | DDC 647.95068/1—dc23/eng/20231017
LC record available at https://lccn.loc.gov/2023041646
LC ebook record available at https://lccn.loc.gov/2023041647

The inside back cover will contain printing identification and country of origin if omitted from this page. In addition, if the ISBN on the back cover differs from the ISBN on this page, the one on the back cover is correct.

SKY10061506_120123

# Contents

# Preface

A foodservice operator's primary responsibility is to deliver quality products and services to their guests at a price mutually agreeable to both parties. In addition, product and service quality must be such that buyers will believe they received excellent value for the money they spent. When this occurs, the operation will prosper, but only if its operating costs have been properly controlled.

The best foodservice operators know that it is wrong to think that "low" costs are good and "high" costs are bad. A foodservice operation with $2 million in sales per year will simply have higher operating costs than a similar operation achieving only $1 million in sales per year. The reason is clear: The amount of products, labor, and equipment needed to sell $2 million worth of food and beverages is greater than that needed to sell only $1 million. If fewer guests are served, there will be lower operating costs, but lower sales and profits as well!

For an effective foodservice operator, then, the question to be considered is not whether costs are high or low. The question is whether costs are *too high* or *too low* given the value a business is creating for its guests. Operating costs must be incurred, and they must be managed in a way that enables the operation to achieve its desired profit levels.

The formula foodservice operators use to calculate their profit level is deceptively easy:

Revenue – Expenses = Profits

An examination of the formula reveals that operators must certainly pay attention to their revenue levels. Effective marketing, quality food and beverage products, and consistently outstanding service will all help to increase the size of this number. But foodservice operators must also pay very close attention to their expenses (costs). Unless costs are properly controlled, profit levels will not be optimized.

The purpose of this book is to teach foodservice operators how they can control their operating costs to ensure their expenses are appropriate for the revenue levels they achieve.

This book is distinctive in that it addresses those areas of cost control that are unique to the foodservice business. Among the many key cost control-related topics addressed in this book are:

✓ The purpose of the Uniform System of Accounts for Restaurants (USAR) when accounting for costs
✓ How to create an accurate sales forecast
  ✓ control costs when purchasing products
  ✓ control costs when receiving products
  ✓ control costs when storing and issuing products
  ✓ control costs in food production
  ✓ control costs in beverage production
  ✓ control costs when serving guests
  ✓ manage service recovery costs
  ✓ manage the cost of labor
  ✓ manage other controllable expenses
  ✓ price menus items for profits
  ✓ develop a revenue security program
  ✓ prepare an accurate operating budget
  ✓ monitor and modify an operating budget

Most importantly, all of these topics and more are addressed in ways specific to foodservice operations of all sizes and in all industry segments.

The ability to properly manage the expenses of a foodservice operation is essential to the operation's success. This is true regardless of whether the operation is a food truck, coffee kiosk, ghost restaurant, quick-service restaurant, full-service operation, or a non-commercial facility. The responsibilities for controlling foodservice costs fall to owners, managers, and production and service staff. Each of these groups plays an important role in the control of costs, and in this book foodservice operators will learn what each of these groups must know, and do, to contribute to the successful operation a foodservice business.

Readers will quickly find that the content of this book is essential to the successful management of their own foodservice operations, and they will also find that the information in each chapter has been carefully prepared to be easy to read, easy to understand, and easy to apply.

# Book Features

In addition to the essential cost control information it contains, special features were carefully crafted to make this learning tool powerful but still easy to use. These features are:

1) **What You Will Learn** To begin each chapter, this very short conceptual bulleted list summarizes key issues readers will know and understand when they complete the chapter.

2) **Operator's Brief** This chapter-opening overview states what information will be addressed in the chapter and why it is important. This element provides readers with a broad summation of all important issues addressed in the chapter.

3) **Chapter Outline** This two-level outline feature makes it quick and easy for readers to find needed information within the body of the chapter.

4) **Key Terms** Professionals in the foodservice industry often use very special terms with very specific meanings. This feature defines important (key) terms so readers will understand and be able to speak a common language as they discuss issues with their colleagues in the foodservice industry. These key terms are also listed at the end of each chapter in the order in which they initially appeared.

5) **Find Out More** In a number of key areas, readers may want to know more detailed information about a specific topic or issue. This useful book feature gives readers specific instructions on how to conduct an Internet search to access that information and why it will be of importance to them.

6) **Technology at Work** Advancements in technology play an increasingly important role in many aspects of foodservice operations. This feature was developed to direct readers to specific technology-related Internet sites that will allow them to see how advancements in technology can assist them in reaching their operating goals.

7) **What Would You Do?** These "mini" case studies located in every chapter of the book take the information presented in the chapter and use it to create a true-to-life foodservice industry scenario. They then ask the reader to think about their own response to that scenario (i.e., *What Would You Do?*).

This element was developed to help heighten a reader's interest and to plainly demonstrate how the information presented in the book relates directly to the practical situations and challenges foodservice operators face in their daily activities.

8) **Operator's 10-Point Tactics for Success Checklist** Each chapter concludes with a checklist of tactics that can be undertaken by readers to improve their operations and/or personal knowledge. For example, in a chapter of the book related to controlling costs in food and beverage production, one point in that chapter's 10-Point Tactics for Success Checklist is:

*Operator can calculate the recipe conversion factor (RCF) required to properly increase or decrease the yield of a standardized recipe.*

# Instructional Resources

This book has been developed to include learning resources for instructors and for students.

## To Instructors

To help instructors (and corporate trainers!) effectively manage their time and enhance student learning opportunities, the following resources are available on the instructor companion website at www.wiley.com/go/hayes/costcontrolfoodservice

- ✓ Instructor's Manual that includes author commentary for "What Would You Do" mini case-study questions.
- ✓ PowerPoint slides for instructional use in emphasizing key concepts within each chapter.
- ✓ A 100-item Test Bank consisting of multiple-choice exam questions, their answers, and the location within the book from which the question was obtained. The test bank is available as a print document and as a Respondus computerized test bank. Note: **Respondus** is an easy-to-use software program for creating and managing exams that can be printed to paper or published directly to Blackboard, WebCT, Desire2Learn, eCollege, ANGEL, and other e-learning systems.

## To Students

Learning how to control costs in foodservice operations will be fun. That's a promise from the authors to you. It is an easy promise to make and keep because working in the foodservice industry is fun. And it is challenging. However, if you work hard and do your best, you will find that you can master all of the important information in this book.

When you do, you will have gained invaluable knowledge that will enhance your skills and help advance your own hospitality career. To help you learn the information in this book, online access to over 200 PowerPoint slides is available to you. These easy-to-read tools are excellent study aids and can help you when taking notes in class.

# Acknowledgments

*Cost Control in Foodservice Operations* has been designed to be the most up-to-date, comprehensive, technically accurate, and reader-friendly learning tool available to those who want to know how to optimize profits by effectively managing costs in a foodservice operation.

The authors thank Catriona King of Wiley for initially working with us to develop the idea for a series of practical books that would help foodservice operators of all sizes more effectively manage their businesses. She was essential in helping conceptualize the need for this book as well as all of the other books in this five-book *Foodservice Operations* series. The five titles in the series are:

✓ *Cost Control in Foodservice Operations*
✓ *Marketing in Foodservice Operations*
✓ *Accounting and Financial Management in Foodservice Operations*
✓ *Managing Employees in Foodservice Operations*
✓ *Successful Management in Foodservice Operations*

We would also like to thank the external reviewers who gave so freely of their time as they provided critical industry and academic input on this series. To our reviewers, Dr. Lea Dopson, Gene Monteagudo, Isabelle Elias, and Peggy Richards Hayes, we are most grateful for your comments, guidance, and insight. Also, thanks to Michael T. Kavanagh, who was a technological friend indeed, when we were most in need!

Books such as this require the efforts of many talented specialists in the publishing field. The authors were extremely fortunate to have Todd Green, Judy Howarth, and Kelly Gomez at Wiley as our publication team. Their efforts went far in helping the authors present the book's material in its best and clearest possible form.

Finally, the authors would like to thank the many students and industry professionals with whom we have interacted over the years. We sincerely hope this book allows us to give back to them as much as they have given to us.

*David K. Hayes, Ph.D.*
*Jack D. Ninemeier, Ph.D.*

# Dedication

*The authors are delighted to have the opportunity to dedicate this book, and this entire* Foodservice Operations *series, to two outstanding and unique individuals.*

## Brother Herman Zaccarelli

*Brother Herman E. Zaccarelli, C.S.C., passed away in 2022 at the Holy Cross House in Notre Dame, Indiana. His professional work included many projects for the hospitality industry, and he published several books and hundreds of articles for numerous trade publications over many years. Among numerous accomplishments, Herman founded Purdue University's Restaurant, Hotel, and Institutional Management Institute in 1976. Later, he served as Director of Business and Entrepreneurial Management at St. Mary's University in Winona, Minnesota.*

*A lifelong learner, at the age of 68, Brother Herman retired to Florida where he earned a bachelor's degree in Educational Administration and a master's degree in Institutional Management.*

*Herman's ideas and concepts have been widely adopted in the hospitality industry, and he assisted many young educators including the authors of this book series. He will be remembered as a colleague with creative ideas who provided significant assistance to those studying and managing in the hospitality industry. Herman was especially helpful in discovering and addressing learning opportunities for Spanish-speaking students, educators, and managers throughout the United States and around the world.*

## Dr. Lea R. Dopson

*A lifelong friend, advisor, and colleague, as well as an outstanding author herself, at the time of her untimely passing in 2022, Lea served as President of the International Council on Hotel, Restaurant, and Institutional Education (ICHRIE) and Dean of the prestigious Collins College of Hospitality Management (Cal Poly Pomona).*

*Lea was a dedicated hospitality professional and a fierce advocate for hospitality students at all levels. Those who knew her were continually in awe of her intelligence and humility.*

*It was especially fitting that Lea was named as a recipient of the H.B. Meek award. That award is named after the individual who started the very first hospitality program in the United States (at Cornell University). Selected annually by the recipient's peers, it goes not to the most outstanding academic professional working in the United States but to the most outstanding academic professional in the entire world. That was Lea.*

*While she is dearly missed, her inspiration goes on everlastingly in the works of the authors.*

# 1

# Cost Control in Foodservice Operations

---

**What You Will Learn**

1) The Importance of Cost Control in a Foodservice Operation
2) Responsibilities for Cost Control
3) The Use of Three Essential Cost Control Tools

---

**Operator's Brief**

In this chapter, you will learn why controlling costs in a foodservice operation is essential to the operation's financial success. The profit formula in a foodservice operation is:

Revenue − Expenses = Profit

In this chapter, you will also learn about the relationship between a food service business's **revenue**, **expenses**, and **profits**. Profit is directly affected by operating costs, and their control is extremely important. The responsibilities for controlling foodservice costs fall to owners, managers, and production and service staff. Each of these groups plays an important role in the control of costs and in this chapter you will learn how and why this is true.

Foodservice operators have a variety of valuable tools available to them. In this chapter, you will learn about three of the most important of them: the operating budget, standardized recipes, and income statements.

Your properly developed operating budget tells what you should expect regarding the generation of revenue, expenses required to obtain that revenue, and the profits likely to result. Standardized recipes are an important tool

*(Continued)*

that helps you ensure that the costs incurred in the production of food and beverage items sold to guests are of the proper quality and cost.

Finally, in this chapter, you will learn that the income statement (also referred to as the profit and loss statement or P&L) is a financial document that summarizes the revenue, expenses, and profit achieved by your operation during a specific time. It serves as a detailed description of your operation's unique "Revenue − Expenses = Profits" formula.

---

**CHAPTER OUTLINE**

The Importance of Cost Control
    The Foodservice Profit Formula
    The Uniform System of Accounts for Restaurants (USAR)
    The Process of Effective Cost Control
Responsibility for Cost Control
    Owner's Responsibilities
    Manager's Responsibilities
    Production Staff Responsibilities
    Service Staff Responsibilities
Essential Cost Control Tools
    The Operating Budget
    Standardized Recipes
    The Income Statement

---

# The Importance of Cost Control

An operator's primary food and beverage responsibility is to deliver quality products and services to their guests at a price mutually agreeable to both parties. In addition, product quality must be such that buyers will believe they received excellent value for the money they spent. As this occurs, the operation will prosper financially.

It is important operators understand that serving guests creates a need for their businesses to incur costs. It is wrong to think that "low" costs are good and "high" costs are bad. A foodservice operation with $2 million in sales per year will have higher costs than a similar operation achieving only $1 million in sales per year. The reason is clear: The amount of products, labor, and equipment needed to sell $2 million worth of food and beverages is greater than that needed to sell only $1 million. If fewer guests are served, there will be lower operating costs, and lower sales and profit as well!

For an effective foodservice operator, the question to be considered is not whether costs are high or low. The question is whether costs are *too high* or *too low* given the value a business is creating for its guests. Costs must be incurred, and they must be managed in a way that enables the operation to achieve desired profit levels.

## The Foodservice Profit Formula

Every foodservice operator must know and understand the components in the profit formula:

$$\text{Revenue} - \text{Expenses} = \text{Profit}$$

When sales are made to guests, a foodservice operation receives revenue, and it also incurs the expenses required to generate those sales. Profit is the amount of money (if any!) that remains after all expenses have been paid. Since it is common in the foodservice industry, the following terms will be used interchangeably: revenues and sales, expenses and costs, and profit and net income.

The profit formula holds true even for managers in the **non-profit sectors** of foodservice.

There are a number of names commonly used to identify non-profit operations. These include non-commercial, institutional, contract feeding, on-site foodservices, and managed services. The number of non-profit foodservice operations is vast and varied and includes food operations in:

K–12 education
Colleges
Universities
Hospitals
Nursing homes
Retirement centers
Correctional facilities
Military facilities
Businesses

**Key Term**

**Revenue:** The term indicating the dollars taken in by a business within a defined time period. Also referred to as "sales."

**Key Term**

**Expenses:** The price paid to obtain the items required to operate a business. Also referred to as "costs."

**Key Term**

**Profit:** The dollars that remain after all of a business's expenses have been paid. Also referred to as "net income."

**Key Term**

**Non-profit sector (foodservice):** Foodservice in an organization where generating food and beverage profits is not the organization's primary purpose. Also referred to as the "non-commercial" sector.

Factories
Private clubs
Sporting arenas and stadiums
National and state parks
Transportation providers including airlines, trains, and cruise ships

The success of a commercial (for-profit) foodservice operation is most often determined by its sales volume (dollars) and profits it generates. However, the success of non-profit operations is often evaluated by participation rate (number of people it serves). In all cases, however, revenue and expenses are just as important to non-profit operations as it is to commercial operations.

Figure 1.1 summarizes how foodservice operators generate profits by using the cash (revenue) they receive from the sale of their products and services.

When the process shown in Figure 1.1 includes the careful control of costs, guest satisfaction and profits can be optimized.

## The Uniform System of Accounts for Restaurants (USAR)

Numerous individuals and groups are interested in a foodservice operation's financial performance. Therefore, this information must be summarized in a manner that is consistent and easily read and understood.

Laws exist that require owners to properly report and pay taxes due, to file certain documents with the government, and to supply accurate business data to various other entities. As a result, many hospitality businesses use a series of standardized (uniform) accounting procedures to report their revenue and expense. These

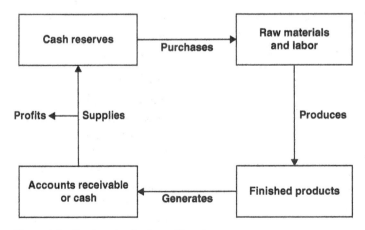

**Figure 1.1** Foodservice Business Flowchart

are called a **uniform system of accounts,** and they represent agreed-upon methods to record financial transactions.

The financial information produced using a uniform system of accounts should be verifiable by a third-party, if needed, and it will be understood by lenders, investors, taxing authorities, and others who must know about this information.

Different foodservice businesses have different accounting needs, and there are uniform systems of accounts produced specifically for individual business segments. In the hospitality industry, some of the best known of these uniform systems are the:

✓ **Uniform System of Accounts for Restaurants (USAR)**[1]
✓ Uniform System of Financial Reporting for Clubs (USFRC)
✓ Uniform System of Accounts for the Lodging Industry (USALI)

**Key Term**

**Uniform system of accounts:** Accounting standards used to help provide accuracy, reliability, uniformity, consistency, and comparability in reporting financial information.

**Key Term**

**Uniform System of Accounts for Restaurants (USAR):** A recommended and standardized (uniform) set of accounting procedures used to categorize and report restaurant revenue and expenses.

Uniform accounting systems are continually reviewed and periodically revised. For example, this book was prepared using reporting principles contained in the eighth and most current edition of the USAR. Specific USAR recommendations will be addressed in detail in relevant portions of this book.

To illustrate why the use of the USAR is so important, assume that an individual owned two Italian restaurants located in two different cities, and both offered the same menu. The owner wants to assess the ability of the two individuals responsible for operating each of the restaurants. It would be very confusing if the two units' managers used different methods for preparing and reporting each of their operations' revenue, expenses, and profits. In this example, unless both operators report and account for their financial performance in a consistent (uniform) manner, the performance of the two operations and their managers could not be properly analyzed and compared.

The use of the USAR to produce accounting information for foodservice operations is not mandatory, but its use is highly recommended. One reason is that a primary purpose of preparing accounting information is to clearly identify revenue, expenses, and profits for a specific time period. The best foodservice

---

1 https://shop.restaurant.org/The-Uniform-System-of-Accounts-for-Restaurants-Print

operators want to do this properly so the financial records of their businesses will accurately reflect their efforts.

Some foodservice operators feel that **accounting** for their businesses is an extremely complex process. While professional accounting does require attention to detail, every foodservice operator can master the basic accounting skills required to effectively record and analyze their revenue, expense, and profits.

**Key Term**

**Accounting:** The system of recording and summarizing a foodservice operation's financial transactions and analyzing, verifying, and reporting the results.

---

**Find Out More**

Some foodservice operators elect to do their accounting in-house without the assistance of outside accounting experts. However, properly accounting for a foodservice operation's revenue and expense is very important. This is especially the case when a foodservice operator seeks a bank loan or investment from outside investors, and their accounting records must reflect the use of the Uniform System of Accounting for Restaurants (USAR).

Professionals working in the foodservice industry have a variety of tools available to assist with their accounting tasks. One of the best is the book "Accounting and Financial Management in Foodservice Operations" written by Drs. David Hayes and Jack Ninemeier and published by John Wiley.

To review the contents of this book prepared for operators who want to utilize the USAR, go to www.wiley.com. When you arrive at the Wiley website, enter "Accounting and Financial Management in Foodservice Operations" in the search bar to examine the outline and content of this valuable accounting resource.

---

### The Process of Effective Cost Control

The effective control of costs is a major responsibility of all foodservice operators and managers. While there are numerous ways to view the cost control process, one method is to consider where the control function lies within the five functions essential to effectively manage any foodservice operation (Figure 1.2):

✓ *Planning:* This initial step in the management process addresses the creation of goals and objectives. It is the first step in the management process because it identifies precisely what the organization wants to achieve through its efforts.

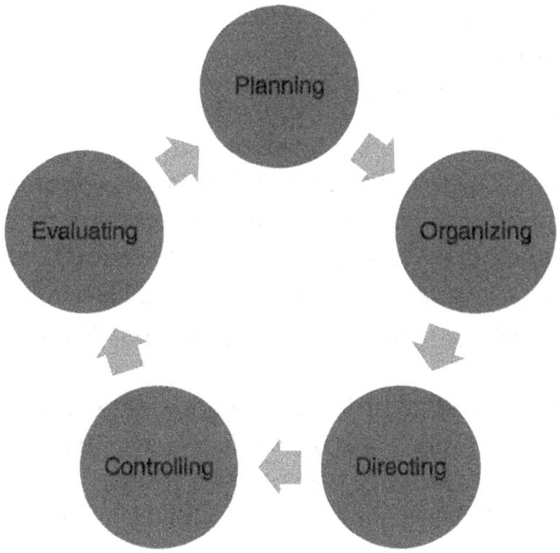

**Figure 1.2** The Management Process

✓ *Organizing:* After its objectives have been identified, an organization must ensure it has the funding, staff, equipment, and raw materials needed to achieve its objectives. These items must then be arranged (organized) to optimize the organization's ability to achieve objectives.

✓ *Directing:* This important management function addresses the task of telling and showing all staff members exactly what is expected of them. When given clear directions, all staff members will know the important roles they play in helping the organization achieve its objectives.

✓ *Controlling:* By continually assessing the work processes and procedures they have put in place, foodservice operators can better identify situations that could prevent their business from meeting its objectives. In the foodservice industry, important processes that must be controlled include those activities related to purchasing, receiving, storing, preparing, and serving menu items.

✓ *Evaluating:* This final management activity requires that an organization assess its current performance and compare the performance to planned or forecasted performance. If significant differences exist, the organization must determine the reason for the differences and then either change objectives and/ or change the methods used to achieve them.

Although this book addresses each of the five key management processes as appropriate, its primary focus is on control. Specifically, it addresses the important things foodservice operators must know and do to properly control costs.

Essentially, the professional control of costs in a foodservice organization requires operators to:

✓ Establish performance standards based on organizational objectives
✓ Measure and report actual performance
✓ Compare actual results to desired (planned for) results
✓ Take corrective actions as needed

An appropriate level of business profits results from effective planning, the use of appropriate management principles, and careful decision making. It is important to understand that profit should not be viewed as what is left over after all bills are paid. In fact, careful planning is necessary to earn a profit, and few investors are attracted to businesses that do not generate enough profit to make their attention worthwhile.

The foodservice business can be very profitable; however, there is no guarantee that an individual operation generates a profit. Some operations do, and others do not. A modification of the profit formula (Revenue − Expenses = Profit) that recognizes and rewards a business owner for the risks associated with ownership is:

Revenue − Desired Profit = Ideal Expense

In this formula, **ideal expense** is an operator's view of the correct or appropriate amount of expense necessary to generate a given level of sales, and **desired profit** is the profit that the owner wants to achieve at that level of revenue.

This modified profit formula clearly places profit as a reward for providing service, not as a "leftover." When foodservice managers deliver quality and value to their guests, anticipated revenue levels and desired profits can be achieved.

Desired profit and ideal expense levels are not, however, easily achieved. It takes a talented foodservice operation team to consistently make the best decisions to optimize revenue while holding expenses to their ideal or appropriate amount. This book was written to teach operators how to make some of these good decisions.

**Key Term**

**Ideal expense:** An operator's view of the correct (appropriate) amount of expense necessary to generate a given quantity of sales.

**Key Term**

**Desired profit:** The profit level that an operator seeks to achieve with a predicted quantity of revenue.

## Responsibility for Cost Control

Regardless of someone's talent, it is seldom possible for a single person to effectively manage all cost control activities in an entire foodservice operation. Instead, properly controlling costs requires a team effort, and each team member has specific responsibilities to address their part of an operation's cost control efforts.

Team members with specific cost control-related responsibilities include:

✓ Owners
✓ Managers
✓ Production staff
✓ Service staff

## Owner's Responsibilities

Not surprisingly, the owners of a foodservice operation will be especially concerned about the proper control of costs. The owners may be one or more individuals, partnerships, or small or very large corporations, but they all will care about the performance of their investment in the operation.

Investors in a foodservice operation generally want to put their money into businesses that will conserve or increase their wealth. Sophisticated owners know that generating significant revenues in a business is important, but only if the revenue is generated with associated costs yielding reasonable profits.

Primary cost-related responsibilities of owners include providing their operating team with the appropriate staff, facilities, equipment, tools, and other resources needed for their jobs. For example, an owner providing a modern **point-of-sales (POS) system** that can track monthly, weekly, daily, and hourly sales enables managers to better control costs related to generating those revenues.

If, however, an owner provides an out-of-date or ineffective POS system (or no POS system at all!), managers will need to spend excessive amounts of time just to record and monitor their sales levels.

**Key Term**

**Point-of-sale (POS) system:** An electronic system that records foodservice customer purchases and payments and other operational data.

## Manager's Responsibilities

On-site managers are generally most responsible for the effective management of costs. Relying on their skills and experience, managers establish performance standards based on the expectations of ownership. The managers must then measure and report the actual performance of their operations and compare that performance to the desired or budgeted (planned for) results.

When significant variations between projected performance and actual performance exist, on-site managers must take corrective actions to ensure their cost control efforts are optimized.

## Production Staff Responsibilities

The efforts of an operation's production staff are incredibly important to the operation's ability to control costs and generate a profit.

An operation's production staff must create the menu items a foodservice operation's guests will purchase. When those products are consistently of high quality and when they satisfy guest expectations, an operation's revenue and profits will increase. If, however, menu items are produced inconsistently or to a low level of quality, revenues and profits will likely suffer.

Even products that are properly prepared must consider the cost guidelines established by managers. For example, if a menu item is sold for $10.00 and its projected cost of production is $3.00, an operation will have $7.00 ($10.00 selling price – $3.00 product cost = $7.00) remaining to pay for its other operating expenses and to contribute to profit levels.

If, however, the same $10.00 menu item that is not properly prepared costs the operation $4.00 to produce, only $6.00 will remain to pay the operation's other operating expenses. The best foodservice production teams know their responsibilities to control costs include the consistent production of tasty menu items at the appropriate cost.

## Service Staff Responsibilities

Foodservice operators sell products, and they deliver services. When a foodservice operation advertises that it offers quality service, good service, or quick service, these concepts directly affect guest perceptions, revenue generation, and the control of costs.

Guests buying from a foodservice operation must put their faith in the operator that the service provided will be of proper quality. An operation's service staff must justify this trust by consistently delivering service at levels that meet (or even exceed!) their guests' expectations.

Most foodservice guests tend to equate the quality of service they receive with the actual person who provides it. As a result, the *rude* waitstaff member will be perceived by guests as providing poor service, while the *cheerful* waitstaff member will be perceived as providing good service (even when the precise tasks performed by the two are similar or even identical).

It is for this reason that many foodservice operators want to hire persons who will directly serve guests for their attitudes—not their initial service skills. The reason: They understand that the real job of a professional food server is to:

✓ Make guests feel welcome
✓ Make guests feel important
✓ Make guests feel special

✓ Make guests feel comfortable
✓ Show a genuine interest if guest expectations were not met and take appropriate corrective actions
✓ Correct any service shortcomings promptly and with a positive attitude

When these goals are achieved, revenue can be optimized, and costs related to the delivery of high-quality service can be controlled. It follows that foodservice operators should pay a great deal of attention to their operations' **front-of-house staff** training and service standardization efforts.

**Key Term**

**Front-of-house staff:** The employees of a foodservice operation whose duties routinely put them in direct contact with guests.

**What Would You Do? 1.1**

"I can't get her attention," said Sharon. "I keep trying, but she isn't looking at us."

Sharon and her co-worker Armando were having lunch at O'Neil's, a small restaurant popular for its soups, salads, and burgers served in a casual setting.

"I know you asked for a side of mayonnaise when you ordered your cheeseburger. She must've forgotten," said Armando.

"I think you're right," said Sharon, "I didn't notice it when she delivered our food. Now I'm not sure about what to do. It's been almost ten minutes, and she hasn't checked back with us."

"And your burger is getting cold," observed Armando. "Maybe they can make you a new one and serve it the right way so it will still be hot."

**Assume you were the manager of O'Neil's. What do you think will be Sharon's assessment of the service and product quality of the burger she received during her visit? If the burger must be replaced, how will that impact the costs you will incur in serving Sharon and Armando's table?**

## Essential Cost Control Tools

This book identifies and explains many tools foodservice operators can use to help manage their costs, and operators can select those tools that best help them achieve financial goals. There are, however, three essential cost control tools that must be utilized by all foodservice operators desiring to effectively manage their costs. These tools are shown in Figure 1.3.

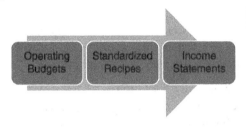

**Figure 1.3** Three Essential Cost Control Tools

## The Operating Budget

An **operating budget** is a financial plan that is very important if a business is to achieve its future financial goals. The operating budget tells foodservice operators what must be done if they are to meet their predetermined sales, costs, and profit objectives. Many operators believe it is the single most important cost control tool available to them.

**Key Term**

**Operating budget:** An estimate of the income and expenses of a business over a defined accounting period. Also referred to as a "financial plan."

While every foodservice operation should approach budgeting from its own perspectives, in most cases an operating budget is produced by:

1) Establishing realistic financial goals
2) Developing an operating budget (financial plan) to achieve the goals
3) Comparing actual operating results with budgeted results
4) Taking corrective action, if needed, to modify operational procedures and/or revise the financial plan to better meet future financial goals

Preparing a budget and staying within its parameters helps operators meet their financial goals. Without such a plan, operators must guess about how much revenue they will generate and how much they should spend to generate those sales.

### Advantages of Operating Budgets

The owners of a foodservice operation want to know what they can expect to earn on their investments, and an operating budget is necessary to project those earnings. Questions related to the amount of revenue likely generated, the amount of cash that will be available for bill payment or distribution as earnings, and the proper timing of major purchases are all important to all foodservice operators. As a result, in organizations of all sizes, proper budgeting is an essential process that must be properly planned and implemented.

The advantages of preparing and using an operating budget are numerous and are summarized in Figure 1.4.

1) It is a way of analyzing alternative courses of action, and it allows operators to examine these alternatives before adopting a specific plan.
2) It requires operators to examine the facts regarding what is necessary to achieve desired profit levels.
3) It provides defined standards that managers can use to develop and enforce appropriate cost control systems.
4) It allows operators to anticipate and prepare for future business conditions.
5) It helps operators periodically carry out a self-evaluation of their business and its progress toward meeting financial objectives.
6) It provides a communication channel in which the operation's objectives can be passed along to various constituencies including owners, investors, managers, and staff.
7) It encourages those who have participated in the preparation of the budget to establish their own operating objectives and evaluation techniques and tools.
8) It provides operators with reasonable estimates of future expense levels and serves as an important aid in determining appropriate selling prices.
9) It identifies time periods in which operational cash flows may need to be supplemented.
10) It communicates realistic financial performance expectations to owners and investors.

**Figure 1.4**   Advantages of Preparing and Using an Operating Budget

**Creating an Operating Budget**

Some operators believe it is not cost-effective to establish an operating budget, and they do not take the time to do so. Creating an operating budget, however, does not need to be a complex process, and similar procedures are needed even if a budget is not planned.

Before operators can begin to develop a budget, they must know about some essentials required for its creation. Note: If these essentials are not addressed before a budget is created, the budget development process is unlikely to yield a budget that is accurate or helpful.

Before creating an operating budget, its developers will need to know about and understand:

1) Prior-period operating results (if an existing operation)
2) Assumptions made about the next period's operations
3) Knowledge of the organization's financial objectives

1) Prior-period operating results

The task of budgeting is easier when an operator knows financial results of prior accounting periods. Experienced foodservice operators know that what has occurred in the past is often a good indicator of what is likely to happen in the future. The further back and more detailed an operator can track their historical revenues and expenses, the more accurate a future budget is likely to be.

2) Assumptions about the next period's operations

Evaluating future conditions and business activity is always a key part of developing an operating budget. Examples of this include the opening of new competitive restaurants in the immediate area, special scheduled occurrences including local sporting events and concerts, and significant changes in operating hours. Local newspapers, trade or business associations, and Chambers of Commerce are possible sources of helpful information about changes in future guest demand.

If significant changes will occur in the operation such as the addition of new menu items and the estimated impact of marketing efforts, assumptions about the impact of these actions are necessary. After important demand-affecting factors have been considered, assumptions regarding revenues and expenses become possible.

3) Knowledge of the organization's financial objectives

An operation's financial objectives may consist of a given profit target that might be a percent of total sales, a total dollar amount, and/or attainment of specific financial and operational ratios. Many of these financial objectives will be determined by an operation's owners based on their desired **return on investment (ROI)**, and the operating budget must address these goals.

**Key Term**

**Return on investment (ROI):** A measure of the ability of an investment to generate income

An investor's ROI is expressed as a percentage of money earned to money invested using the following formula:

$$\frac{\text{Money Earned on Funds Invested}}{\text{Funds Invested}} = \text{ROI}$$

For example, if an operator invested $1,000,000 in a foodservice business and generated $200,000 in profits during the first year, the operator's ROI for that year would be calculated as

$$\frac{\$200,000}{\$1,000,000} = 0.20 \text{ or } 20\% \text{ ROI}$$

The budgeted profit level and resulting ROI an operation seeks can be achieved when the operation realizes its budgeted sales levels, and it spends only what was budgeted to generate those sales. If revenues fall short of forecast and/or if expenses are not reduced to match the shortfall, budgeted profit levels are not likely to be achieved.

Foodservice operators establish their budgets by focusing on each of the three components of the profit formula addressed earlier in this chapter:

✓ Revenue
✓ Expenses
✓ Profits

### Revenue

Accurately forecasting an operation's sales (see Chapter 2) is critical because all forecasted expenses and profits will be based on revenue forecasts. Effective operators know that, in most cases, revenues should be estimated on a monthly (or weekly) basis and then be combined to create an annual operating budget.

Forecasting revenues to establish the revenue portion of an operating budget is not an exact science. However, it can be made more accurate if operators:

✓ **Review historical revenue records.** Many operators begin the revenue forecasting process by reviewing revenue records from previous years. A budget being planned for one or more months can often yield trends or data that can be used to predict future revenue patterns. This is generally useful because historical revenue "numbers" are often used to develop future budgets.
✓ **Consider internal factors affecting revenues.** In this step, operators consider any significant changes in the type, quantity, and direction of their marketing efforts. Other internal activities that can impact future revenues include those related to facility renovation that may affect capacity and times when the operation may be disrupted because of renovations. In other operations, the number of hours to be open or menu prices may change. Any internal initiation or change that an operator believes will likely impact future revenues should be considered in this step.
✓ **Consider external factors affecting revenues.** There are numerous external issues that could affect an operation's revenue forecasts. These include new competitors' planned openings (or competitors' closings) and other factors such as road improvements or construction that could disrupt normal traffic patterns. Other external factors may include changes in local regulations regarding signage and, important for many hospitality businesses, economic upturns or downturns that impact potential guests as they spend discretionary income on hospitality services.

> **Key Term**
>
> **Fixed cost:** An expense that stays constant despite increases or decreases in sales volume.
>
> **Key Term**
>
> **Variable cost:** An expense that generally increases as sales volume increases and decreases as sales volume decreases.

### Expenses

When preparing budgets, operators must account for each **fixed cost** and **variable cost** they will incur in generating revenue.

Fixed costs are simple to forecast because items such as mortgage payments or rent typically stay the same from month-to-month.

Variable costs, however, are directly related to the amount of revenue produced by a foodservice operation. An operation that forecasts sales of 50 prime rib dinners on Friday night will have meat and server costs higher than the operator in a similar facility that forecasts the sale of only 10 such dinners. Variable expenses such as food and beverage and labor costs are directly affected by sales levels. Fixed and variable costs can be budgeted by dollar amount or as a percentage of sales.

### Profits

The profit formula states that after expenses are subtracted from revenue, the remaining amount is profit. For example, if an operator budgets revenues of $50,000 in a specific month and expenses of $45,000 in that month, a $5,000 profit will result:

$$\$50,000 \text{ budgeted revenue} - \$45,000 \text{ budgeted expenses} = \$5,000 \text{ budgeted profit}$$

When an operation's expenses exceed its revenue, however, no profits are generated. Instead, the operation budgets for a loss. This can be the case, for example, when an operation initially opens, or when other factors dictate that expenses will exceed revenue for a specific time.

---

**Technology at Work**

The preparation and monitoring of operating budgets in the foodservice industry can be a time-consuming process. Today, however, there are a variety of software programs available, in both PC and Mac formats, to help owners of foodservice businesses develop and monitor their operating budgets.

To review the features of some of these helpful tools, enter "budgeting software for restaurants" in your favorite search engine and view the results.

---

### Standardized Recipes

Experienced operators know that, to stay within their budgets, they must be able to predict and control their costs. Although it is the menu that determines what is to be sold, and at what price, the **standardized recipe** (addressed in greater detail in Chapter 6) controls both the quantity and the quality of what a kitchen or bar will produce.

**Key Term**

**Standardized recipe:** The instructions needed to consistently prepare a specified quantity of food or drink at an expected quality level.

A standardized recipe details the procedures used to prepare and serve each of an operation's food or beverage items. A standardized recipe ensures that each time guests order a menu item they receive exactly what management intended them to receive. A high-quality standardized recipe contains the following information:

1) Menu item name
2) Total recipe yield (number of portions produced)
3) Portion size
4) Ingredient list
5) Preparation/method discussion, if needed
6) Cooking time and temperature
7) Special instructions, if necessary
8) Recipe cost (optional)

Figure 1.5 shows a sample standardized recipe for an operation. If this standardized recipe represents the quality and quantity the operation wishes its guests to receive, and if it is followed carefully each time the item is prepared, then guests will always receive the quality and value the operation intended.

**Roast Chicken**

| | |
|---|---|
| Special Instructions: | Recipe Yield: |
| Serve with | 48 portions |
| Crabapple Garnish (see Crabapple Garnish Standardized Recipe) | Portion Size: ¼ chicken |

Serve on 10-in. plate.

| Ingredients | Amount | Method |
|---|---|---|
| Chicken Quarters (twelve 3–3½-lb. chickens) | 48 ea. | Step 1. Wash chicken; check for pinfeathers; place on 24-inch × 20-inch baking sheets. |
| Butter (melted) | 1 lb. 4 oz. | Step 2. Clarify butter; brush liberally on chicken quarters; combine all seasonings and mix well; sprinkle over chicken quarters. |
| Salt | ¼ C | |
| Pepper | 2 T | |
| Paprika | 3 T | |
| Poultry Seasoning | 2 t | |
| Ginger | 1½ t | |
| Garlic Powder | 1 T | Step 3. Roast in oven at 325°F for 1.5 hours or to an internal temperature of at least 165°F. |

**Figure 1.5** Standardized Recipe: Roast Chicken

The best foodservice operators create and then enforce the use of standardized recipes because:

1) Accurate purchasing is impossible without the existence and use of standardized recipes.
2) Dietary concerns require some foodservice operators to know the exact ingredients and the correct amount of nutrients in each serving of a menu item.
3) Accuracy in menu laws require that foodservice operators can tell guests about the type and amount of ingredients in the recipes.
4) The responsible service of alcohol is impossible without adhering to appropriate portion sizes in standardized drink recipes.
5) Accurate recipe costing and menu pricing is impossible without standardized recipes.
6) Matching food and beverages prepared and sold is impossible without standardized recipes.
7) New production employees can be trained faster and better with standardized recipes.
8) The computerization of a foodservice operation is impossible unless the elements of standardized recipes are in place. In fact, the advantages of using advanced technological tools are restricted or even eliminated if standardized recipes are not used.

Experienced foodservice operators agree that use of standardized recipes is essential to any serious effort to produce consistent, high-quality food and beverage products at a known cost to stay within cost budgets.

## The Income Statement

Operating budgets estimate income and expenses, but a foodservice operation's *actual* revenue, expense, and profits for a particular time period are summarized and reported on its **income statement**.

The income statement precisely reports a business's actual revenues from all its revenue-producing sources, the expenses required to generate all revenues, and the business's resulting profits or losses that are often referred to as the operation's **net income**.

**Key Term**

**Income statement:** Formally known as "The Statement of Income and Expense," a report summarizing a foodservice operation's profitability including details regarding revenue, expenses, and profit (or loss) incurred during a specific accounting period. Also commonly called the profit and loss (P&L) statement.

**Key Term**

**Net income:** A number calculated as revenue minus operating expenses including depreciation, interest, and taxes, among others. It is useful to assess how much revenue exceeds the expenses of operating a business in a defined time period.

There are several ways foodservice operators *could* report their revenue, expense, and profits. The USAR has recommendations for how an income statement *should* be produced. The USAR recommendations are not mandatory, but they are highly recommended and are followed by most operators and all hospitality accounting firms.

To illustrate the production of an income statement using USAR recommendations, consider Figure 1.6. It shows the USAR-suggested income statement format used to report revenue and expenses for Shondra's Restaurant for the year ending December 31, 20xx. Note: The "Line" column has been added by the authors for the reader's ease in identifying data locations.

The USAR income statement is arranged on the above income statement *from the expenses that are most controllable to least controllable* by a foodservice operator. The format of a USAR income statement can best be understood by dividing it into its main sections:

1) Sales (Revenue)
2) Total Cost of Sales
3) Total Labor
4) Prime Cost
5) Other Controllable Expenses
6) Non-Controllable Expenses
7) Income Before Income Taxes (Profits)

### Sales (Revenue)

After listing the name of an operation and the accounting period addressed, the USAR format for an income statement lists sales (Line 1; revenue, first and on the top line). **Topline revenue** represents the total sales generated by a business before any expense deductions.

Historically, food sales (Line 2) include income from the sale of all food items and non-alcoholic beverages such as soft drinks, coffee, tea, milk, bottled water, and fruit juices. When alcoholic beverages are sold, the revenue from the sales of these beverage products (Line 3) is added to food sales to yield an operation's **total sales** (Line 4).

**Key Term**

**Topline revenue:** Sales or revenue shown on the top of a business's income statements.

**Key Term**

**Total sales:** The sum of food sales and alcoholic beverage sales generated in a foodservice operation.

When using the USAR, the physical layout of the income statement for different types of business can vary somewhat in their formats. For example, a foodservice operation that serves alcohol would want its income statement to identify the

**Shondra's Restaurant**

Income Statement

For the Year Ended December 31, 20xx

| Line | | |
|------|---|---|
| 1 | **SALES** | |
| 2 | Food | $1,891,011 |
| 3 | Beverage | $ 415,099 |
| 4 | **Total Sales** | **$2,306,110** |
| | | |
| 5 | **COST OF SALES** | |
| 6 | Food | $ 712,587 |
| 7 | Beverage | $ 94,550 |
| 8 | **Total Cost of Sales** | $ 807,137 |
| | | |
| 9 | **LABOR** | |
| 10 | Management | $ 128,219 |
| 11 | Staff | $ 512,880 |
| 12 | Employee Benefits | $ 99,163 |
| 13 | **Total Labor** | $ 740,262 |
| | | |
| 14 | **PRIME COST** | **$1,547,399** |
| | | |
| 15 | **OTHER CONTROLLABLE EXPENSES** | |
| 16 | Direct Operating Expenses | $ 122,224 |
| 17 | Music and Entertainment | $ 2,306 |
| 18 | Marketing | $ 43,816 |
| 19 | Utilities | $ 73,796 |
| 20 | General and Administrative Expenses | $ 66,877 |
| 21 | Repairs and Maintenance | $ 34,592 |
| 22 | **Total Other Controllable Expenses** | $ 343,611 |
| | | |
| 23 | **CONTROLLABLE INCOME** | $ 415,100 |
| | | |
| 24 | **NON-CONTROLLABLE EXPENSES** | |
| 25 | Occupancy Costs | $ 120,000 |
| 26 | Equipment Leases | $ 0 |
| 27 | Depreciation and Amortization | $ 41,510 |
| 28 | **Total Non-Controllable Expenses** | $ 161,510 |
| | | |
| 29 | **RESTAURANT OPERATING INCOME** | $ 253,590 |
| | | |
| 30 | Interest Expense | $ 86,750 |
| | | |
| 31 | **INCOME BEFORE INCOME TAXES** | $ 166,840 |

**Figure 1.6** Sample USAR Income Statement

revenue generation and costs associated with serving alcoholic drinks. In a family-style pancake restaurant that serves only food and non-alcoholic beverages, the income statement would not include a line for alcoholic beverages.

The USAR's suggested income statement format recommends, at minimum, that food and (alcoholic) beverage sales be reported separately. There are a variety of reasons for this recommendation. They include the ability to better control costs when these sales are separated, and the requirement of most states to record alcohol sales separately from food sales.

In Figure 1.6, Shondra's food sales are separated from alcoholic beverage sales. In other operations, individual revenue categories may be created for catering sales, on-premises banquet sales, take-out versus dine-in sales, merchandise (e.g., logo hats, cups, and T-shirts), or any other revenue category that would be helpful to the operation's managers as they analyze their sales.

In some foodservice operations, the sale of logoed and other merchandise types can be significant. Those readers who are familiar with the Cracker Barrel restaurant group are likely aware of the significant amount of merchandise it sells. Note: Even if an operation's merchandise sales are relatively small, they should be reported separately on the income statement.

---

**Find Out More**

To achieve desired sales levels, foodservice operators must have an effective marketing plan in place. Increasingly, an operation's marketing efforts must include both traditional approaches and newer Internet-based approaches.

While there are a large number of publications addressing general marketing strategies for businesses, few publications exclusively address the marketing needs of foodservice operations. One of the best and most up-to-date marketing resources available to foodservice operators that exclusively addresses the marketing of foodservice operations is *Marketing in Foodservice Operations* (January 2024) by Drs. David K. Hayes and Jack D. Ninemeier.

To learn more about the content and availability of this valuable new publication, enter Wiley: *Marketing in Foodservice Operations* in your favorite search engine and review the results.

---

## Total Cost of Sales

Food and beverage **cost of sales** (Line 5) are entered separately on the income statement. The reporting of food sales and the related cost of food sales (Line 6) separately from the cost of

### Key Term

**Cost of sales:** The total cost of the products used to make the menu items sold by a foodservice operation.

beverage sales (Line 7) is useful for analyzing a foodservice operation. Without this separation an inefficient food operation could be covered up by a highly profitable beverage operation (or the reverse could occur).

A separate income statement listing for "Cost of Sales: Food" and "Cost of Sales: Beverages" equal to total cost of sales (Line 8) is integral to product control procedures. This topic will be discussed more fully in Chapter 6 (Cost Control in Food and Beverage Production).

## Labor

While **payroll** generally refers to salaries and wages a foodservice operation pays to its employees, the USAR income statement provides greater labor cost-related details.

"Payroll" as shown in Figure 1.6, "Labor" (Line 9), for a USAR-formatted income statement is separated into three categories:

### Key Term

**Payroll:** The term commonly used to indicate the amount spent for labor in a foodservice operation. Used for example in "Last month our total payroll was $28,000."

✓ Management
✓ Staff
✓ Employee Benefits

Management (Line 10) includes the total amount of salaries paid during the accounting period, and Staff (Line 11) refers to payments made to hourly (non-salaried) workers.

The employee benefits category (Line 12) includes the cost of all benefits payments made for managers and hourly workers. Some benefit payments are mandatory (such as FICA [Social Security]), and others are voluntary (such as the cost of providing health insurance).

Specific employee benefits paid by an operation will vary but can include:

✓ FICA (Social Security) taxes including taxes due on employees' tip income
✓ FUTA (federal unemployment taxes)
✓ State unemployment taxes
✓ Workers' compensation
✓ Group life insurance
✓ Health insurance, including:
  - Medical
  - Dental
  - Vision
  - Disability

✓ Pension/retirement plan payments
✓ Employee meals
✓ Employee training expenses
✓ Employee transportation costs
✓ Employee uniforms, housing, and other benefits
✓ Vacation/sick leave/personal days
✓ Tuition reimbursement programs
✓ Employee incentives and bonuses

Not every operation will incur all the benefit costs listed above, but some operations may have all of these and perhaps even additional benefits.

**Total labor** expense (Line 13) is listed prominently on a USAR income statement because controlling and evaluating total labor cost is important in every foodservice operation. In fact, many operators feel it is even more important to control labor costs than product costs. One reason is that labor and labor-related benefit costs comprise a larger portion of their operating costs than do food and beverage product costs.

**Key Term**

**Total labor:** The cost of the management, staff, and employee benefits expense required to operate a business.

## Prime Cost

**Prime cost** (Line 14) is defined as an operation's total cost of sales added to its total labor expense. Note: As previously defined, "total cost of sales" represents the amount paid for the food and beverage products sold by an operation. In contrast, "total labor" represents the cost of the management, staff, and employee benefits expense required to operate the business.

**Key Term**

**Prime cost:** An operation's cost of sales *plus* its total labor costs.

Prime cost is clearly listed on the income statement because it is an excellent indicator of an operator's ability to control product costs (cost of sales) and labor costs, the two largest expenses in most foodservice operations.

The prime cost concept is also important because, when prime costs are excessively high, it is difficult to generate a sufficient level of profit in a foodservice operation even when other controllable and non-controllable expenses are well-controlled.

While each foodservice operation is different, prime costs in the range of 60% to 70% of revenue are common for **full-service restaurant**

**Key Term**

**Full-service restaurant:** A foodservice operation at which servers deliver food and drink offered from a printed menu to guests seated at tables and/or booths.

operations, while prime costs below 60% are most common in **fast casual** and **quick-service restaurants (QSR)**.

## Other Controllable Expenses

As shown in Figure 1.6, after prime cost, the next major section is reserved for identifying **other controllable expenses** (Line 15): the non-food/non-alcoholic and labor costs controlled by managers and incurred from operating the business. In most cases, other controllable expenses can be influenced by a foodservice operator's own decisions. The USAR identifies and allows for the use of several "other controllable expense" categories.

The specific other controllable expense categories of an operation listed on its income statement may vary. As shown in Figure 1.6, in the example, Shondra's Restaurant's controllable expenses include:

**Direct Operating Expenses** (Line 16): Operators use this expense category to list the cost of uniforms, laundry, supplies, menus, kitchen tools, and other items incidental to service in dining areas to provide support in kitchen and storage areas.

**Music and Entertainment** (Line 17): These costs, if significant, should be shown separately. Many foodservice operations offer little or no music or entertainment. When this is the case and the expenses are small, they may be recorded as "Miscellaneous" within the direct operating expense category.

**Marketing** (Line 18): This expense category includes newspaper, magazine, radio and TV, and Internet advertising, and expenses for outdoor signs and direct mailings. Loyalty program costs, donations, and special events that promote the business can be additional marketing costs. In some cases, marketing fees associated with third-party meal-delivery companies such as DoorDash and Uber Eats are included in this category.

**Key Term**

**Fast casual (restaurant):** A sit-down foodservice operation with no wait staff or table service. Customers typically order off a menu board and seat themselves or take the purchased food elsewhere.

**Key Term**

**Quick-service restaurant (QSR):** Foodservice operations that typically have limited menus and often include a counter at which customers can order and pick up their food. Most quick-service restaurants also have one or more drive-through lanes that allow customers to purchase menu items without leaving their vehicles.

**Key Term**

**Other controllable expenses:** Expenses (costs) that a foodservice operator can influence with increases or decreases based on business decisions. Examples include marketing costs and utility costs.

**Utilities** (Line 19): This category includes the cost of water, sewage, electricity, and gas used to heat the building. When facilities are rented and the restaurant pays the utilities, these costs are recorded as "utilities" rather than rent.

**General and Administrative Expenses** (Line 20): This group of costs includes expenses generally classified as operating "overhead." These expenses are necessary for business operation as opposed to being directly connected with serving guests. They often include the cost of items such as office supplies, postage, credit card fees, telephone charges, data processing, general insurance, professional fees, and security services.

**Repairs and Maintenance** (Line 21): These expenses include painting and decorating costs, maintenance contracts for elevators and machines, and repairs to an operation's equipment and mechanical systems. This section is not used to record the purchase of new equipment.

The individual entries in Other Controllable Expenses are summed to yield the amount for Total Other Controllable Expenses amount (Line 22). That amount and the operation's prime costs are subtracted from sales to yield the operation's **Controllable Income** (Line 23).

Controllable income is often used to evaluate the effectiveness of unit-level managers because it represents sales minus only those costs that unit managers generally can directly control or influence. In many foodservice companies, controllable income is the basis for determining at least some portion of a foodservice manager's incentive (bonus) pay.

**Key Term**

**Controllable income:** The amount of revenue remaining after an operation's prime costs and other controllable expenses have been subtracted from its total sales.

## Non-Controllable Expenses

**Non-controllable expenses** (Line 24) are listed next on the income statement in Figure 1.6, and they include all costs not under the immediate control of management.

Non-controllable expenses include those previously committed and must be incurred irrespective of an operation's sales volume. In a foodservice operation, examples of non-controllable expenses include rental payments, interest on long-term loans, and **depreciation**.

Non-controllable expenses listed in Figure 1.6 include:

**Key Term**

**Non-controllable expenses:** Costs that, in the short run, cannot be avoided or altered by management decisions. Examples include lease payments and depreciation.

**Key Term**

**Depreciation:** The allocation of equipment costs and other depreciable assets based on the projected length of their useful life.

**Occupancy costs** (Line 25): These include the costs of renting buildings and land, property taxes, and insurance. These expenses can vary considerably between foodservice operations. Since the owners of a business most often control these costs, they are only infrequently under the direct control of those operating the business.

**Equipment leases** (Line 26): These expenses (if any) include the costs incurred from the leasing or renting of equipment used in a foodservice operation. Common examples include charges for leasing POS systems, beverage dispensing equipment, and ice machines. In the example shown in Figure 1.6, no equipment lease cost is incurred.

**Depreciation and Amortization** (Line 27): These **non-cash expenses** are the result of depreciating buildings and **furniture, fixtures, and equipment (FF&E)**.

The individual entries in the Non-Controllable Expenses section of the income statement are summed to yield Total Non-Controllable Expenses (Line 28). That amount is then subtracted from Controllable Income to yield the operation's Restaurant Operating Income (Line 29).

Restaurant operating income represents all of an operation's revenue minus all of its controllable and non-controllable expenses.

**Key Term**

**Occupancy costs:** Costs related to occupying a space including rent, real estate taxes, personal property taxes, and insurance on a building and its contents.

**Key Term**

**Non-cash expense:** Expenses recorded on the income statement that do not involve an actual cash transaction. Examples include depreciation and amortization, which are expenses, and an income statement entry reduces operating income without a cash payment.

**Key Term**

**Furniture, fixtures, and equipment (FF&E):** Movable furniture, fixtures, or other equipment that have no permanent connection to a building's structure.

### Income Before Income Taxes (Profit)

It is interesting to note that the word "profit" does not actually appear anywhere on a USAR income statement. Some foodservice operators consider Restaurant Operating Income (Line 29) to be their business's profit because it represents an operation's sales (revenue) minus all controllable and non-controllable expenses, and it reflects the profit formula introduced earlier in this chapter:

Revenue – Expenses = Profit

Restaurant Operating Income in Line 29 does not consider any **interest expense** payments made by a business. Interest Expense (Line 30) is the cost of borrowing money and is recorded on this line of the income statement even if the interest incurred has not yet been paid.

Restaurant operating income represents all of an operation's revenue minus its controllable and non-controllable expenses, and it may be further adjusted to account for corporate overhead, interest expense, or other owner-controlled expenses.

**Income Before Income Taxes** (Line 31) on the income statement is calculated as Restaurant Operating Income minus Interest Expense.

In Figure 1.6, the operation's Income Before Income Taxes is a number that is frequently referred to as a business's "profit." If total operating expenses *exceed* revenue, the resulting negative numbers (losses) shown on this line of an income statement are typically designated in one of the three ways:

**Key Term**

**Interest expense:** The cost of borrowing money.

**Key Term**

**Income before income taxes:** The amount of money remaining after an operation's interest expense is subtracted from the amount of its restaurant operating income. Also, a business's profit before paying any income taxes due on the profits.

1) By a minus "−" sign in front of the number. For example, a $1,000 loss would be shown in the budget as −$1,000.
2) By brackets "( )" around the number. For example, a $1,000 loss would be presented as ($1,000).
3) With red ink rather than black ink to designate the loss amount. For example, a $1,000 loss would be presented in the budget as "$1,000," but the number would be printed in red. Note: This approach gives rise to the slang phrase to "operate in the red": a business that is not making a profit. In a similar vein, to "operate in the black" indicates the business is profitable.

The income statement is a valuable cost control tool because it allows operators to compare their budgeted (forecasted) results with their actual results. Significant differences between the two can then be investigated and addressed.

A food service operation's expenses are directly tied to the revenue they achieve. Therefore, it is essential that operators concerned with controlling costs effectively forecast their future sales. When they do, they can better forecast their future expenses. Accurately forecasting future sales in a food service operation is so important it will be the sole topic of the next chapter.

---

**Technology at Work**

The demands of operating a foodservice business are such that, in many cases, foodservice operators employ a professional accountant whose responsibilities include ensuring proper recording of income and expenses and the creation of the operators' income statements.

Some foodservice operators, however, prefer to do their own accounting, and they utilize accounting software developed specifically for the foodservice industry. QuickBooks, Restaurant 365, and ZipBooks are examples of accounting software packages used by foodservice operators.

To learn more about the features and costs of some currently popular foodservice-specific accounting software programs, enter "accounting packages for restaurants" in your favorite search engine and review the results.

---

**What Would You Do? 1.2**

"Don't keep me in suspense," said Lani, "how did we do?"

Lani, the general manager of Reggie's Pizza Parlor, was talking to Reggie Stone, the restaurant's owner. Reggie had just been e-mailed the previous month's P&L from their accountant.

"Well, the good news," said Reggie, "is that our sales exceeded our forecast. I guess maybe the lesser good news is that our expenses exceeded forecasts as well!"

The accountant for Reggie's Pizza Parlor used the Uniform System of Accounts for Restaurants (USAR) to prepare the business's monthly P&L, and it usually took the accountant a few days to finalize the previous month's numbers. It was the third of the month and the previous month's P&L had just been received.

One important part of Lani's job was to compare the business's actual performance to its forecasted performance.

"Forward that e-mail to me," said Lani, "Let me take close look at those expenses and our profits."

**Assume you were Lani. Since sales exceeded the forecasts, would you be surprised to find that expenses exceeded the forecast as well? How do you think your budgeted profits would be affected when your business exceeds its sales forecast, but also exceeds its expense forecast?**

## Key Terms

| | | |
|---|---|---|
| Revenue | Operating budget | Quick-service |
| Expenses | Return on investment | restaurant (QSR) |
| Profit | (ROI) | Other controllable |
| Non-profit sector | Fixed cost | expenses |
| (foodservice) | Variable cost | Controllable income |
| Uniform system of | Standardized recipe | Non-controllable |
| accounts | Income statement | expenses |
| Uniform System of | Net income | Depreciation |
| Accounts for | Topline revenue | Occupancy costs |
| Restaurants (USAR) | Total sales | Non-cash expense |
| Accounting | Cost of sales | Furniture, fixtures, and |
| Ideal expense | Payroll | equipment (FF&E) |
| Desired profit | Total labor | Interest expense |
| Point-of-sales | Prime cost | Income before |
| (POS) system | Full-service restaurant | income taxes |
| Front-of-house staff | Fast casual (restaurant) | |

### Operator's 10-Point Tactics for Success Checklist

Evaluate your need for, and the current status of, each of the following operational tactics. For those tactics you think are important, but not yet in place, develop an action plan for its implementation including who will be responsible for the tactic's completion and the target date by which it should be completed.

| | | | | If Not Done | |
|---|---|---|---|---|---|
| Tactic | Don't Agree (Not Done) | Agree (Done) | Agree (Not Done) | Who Is Responsible? | Target Completion Date |
| 1) Operator can identify the three components of the foodservice profit formula. | — | — | — | | |
| 2) Operator discerns the importance of utilizing the Uniform System of Accounts for Restaurants (USAR) when reporting the financial performance of a foodservice operation. | — | — | — | | |

*(Continued)*

| Tactic | Don't Agree (Not Done) | Agree (Done) | Agree (Not Done) | If Not Done | |
|---|---|---|---|---|---|
| | | | | Who Is Responsible? | Target Completion Date |
| 3) Operator understands that the effective control of costs in a foodservice operation is an ongoing process. | —— | —— | —— | | |
| 4) Operator recognizes the important role of business owners in the effective management of costs. | —— | —— | —— | | |
| 5) Operator recognizes the important role of managers in the effective management of costs. | —— | —— | —— | | |
| 6) Operator recognizes the important role of production staff in the effective management of costs. | —— | —— | —— | | |
| 7) Operator recognizes the important role of service staff in the effective management of costs. | —— | —— | —— | | |
| 8) Operator understands the importance of an operating budget in the effective management of costs. | —— | —— | —— | | |
| 9) Operator understands the importance of utilizing standardized recipes in the effective management of costs. | —— | —— | —— | | |
| 10) Operator understands the importance of and the components of the income statement (P&L) to effectively manage costs. | —— | —— | —— | | |

# 2

# Sales Forecasts

**What You Will Learn**

1) How to Properly Record Current Sales
2) How to Calculate Percentage Increases or Decreases in Sales
3) How to Accurately Predict Future Sales

**Operator's Brief**

To best control operating costs, an operator should have a good idea of how many guests will be served on any given day in the future. In this chapter, you will learn that having a record of previous sales (sales histories) is an important tool when forecasting your future sales. The reason: Future sales in your operation can be estimated, in part, based on previous sales.

The dollar amount of previous sales can be recorded as either a fixed average, or a rolling average and, in this chapter, you will learn how to utilize both averaging tools.

In addition to knowing how many guests you are likely to serve in the future, it is important to know how much each guest is likely to spend. The ability to calculate the average sale per guest is important, and you will learn how this is done.

In many cases you will find that there are changes in the sales levels you achieve over time. These sales variances are important to understand, and they can be calculated as either a dollar amount or as a percentage of previous sales. In this chapter, you will learn how to calculate both types of variances.

The accurate forecasting of future sales will enable you to have the needed food and beverage products and staff available to serve your incoming guests. You can forecast your future sales by dollar amount, by the number of guests to be served, and/or by average sale per guest. In this chapter, you will learn how to make sales forecasts based on each of these three important factors.

# The Importance of Accurate Sales Forecasts

The development of an accurate operating budget described in the previous chapter begins with an estimate of topline revenue (see Chapter 1). This forecast of future sales is extremely important because, when opening their property's doors at the beginning of the day, the question most foodservice operators ask themselves is very simple: "How many guests will I serve today?"

The answer to this question is critical because the guests served by an operation provides the revenue needed to pay operating expenses and create a profit. If too few guests are served, total revenue may not equal the operation's costs, even if those costs are well-managed.

In addition, decisions regarding the type and quantity of food and beverage products to purchase depend on estimating the number of guests who will purchase these items. The labor required to serve guests is also determined based on an operator's "best guess" of the projected number of guests to be served and what these guests will buy.

In an ongoing operation, future sales estimates (projected sales) can be heavily based on an operation's **sales history**. Experienced foodservice operators know that what has happened in the past is often a good predictor of what will occur in the future. Likewise, the operators who can best predict the future are those who are most prepared to control how it will affect them.

**Key Term**

**Sales history:** A record of the number of guests served and the revenues generated in a defined time period.

In the hospitality industry, there are several ways of counting (defining) sales. In the simplest case, sales can be defined as the amount of revenue dollars (or other currency) generated during a predetermined time period. The time may be defined as an hour, a shift, a day, a week, a month, or a year. When used this way, sales and revenue are considered interchangeable terms.

Foodservice operators can record their actual sales for a current time period in several ways. Most foodservice managers utilize a computerized point of sale (POS) system (see Chapter 1), which, among other functions, records sales, and payment information.

Alternatively, manually produced guest checks or head counts are other methods foodservice operators might use to identify the amount of sales generated. Today, however, even the smallest of foodservice operations should take advantage of the speed and accuracy provided by modern POS systems when recording their operations' current sales.

When operators estimate the *future* number of guests they will serve and the revenue to be achieved in a given time period, they will have created a **sales forecast**. Accurate sales histories are an important tool when operators produce their sales forecasts.

**Key Term**

**Sales forecast:** A prediction of the number of guests to be served and the revenues to be generated in a defined, future time period.

It is important to know that a distinction is made in the foodservice industry between actual sales (revenue) and **sales volume**.

**Key Term**

**Sales volume:** The number of units sold.

To illustrate the difference between sales and sales volume, consider Tamar, a bagel shop operator whose Monday business consisted of $3,000 in sales (revenue) because she sold 1,000 bagels (sales volume) at $3.00 each. It is important for Tamar to know how much revenue was taken in. With this knowledge she can evaluate the expenses required to generate the revenue that she produced based on the number of units (bagels) that were sold. With this information, she can be better prepared to serve additional guests the next day.

In many non-profit areas of the hospitality industry such as in retirement centers and college and university residence halls, there may be no cash or little cash that changes hands during a specific meal period. In these situations, consumers often use an account number to identify the source of product purchases. However, the operators of these facilities still create sales information during meal periods, and they would be interested in knowing sales volume (the amount of food consumed by the residents or students on any specific day). This is critical information because all foodservice operators must be prepared to answer the questions, "How many individuals did I serve today, and how many should I expect tomorrow?" In some cases, a foodservice operation may generate a blend of cash and noncash sales and this information is equally as important.

Consider Zafira Barzani, a hospital foodservice director. It is very likely that Zafira will be involved in serving both cash-paying guests (in her open-to-the-public cafeteria) and noncash patients (with tray-line-assembled meals that are delivered to patients' rooms). In addition, meals for hospital employees may be made as cash sales, but at a reduced or subsidized rate. Zafira's operation will create sales information each day, and it will be important for her and her staff to know, as accurately as possible, how many of each type of guest she will serve and the menu items these guests will be served.

An understanding of anticipated sales (revenue dollars and/or guest counts) helps operators to plan for the right number of staff members working with the right amounts of available products at the right time. In this way, they can begin to effectively manage their costs. In addition, to the use of accurate sales records for purchasing and staffing, sales records are also valuable to an operator developing labor standards to improve efficiency. Consider, for example, the operator managing a large restaurant with 480 seats. If an individual server can serve 40 guests at lunch, the operator would need 12 servers (480 seats/40 guests per server = 12 servers) per lunch shift if they forecasted that all seats were filled. If the operator keeps no accurate sales histories and makes no sales forecasts, too few or too many servers might be scheduled to work.

Using accurate sales records, a sales history can be developed for each foodservice outlet an operator manages, and better operating decisions can be made when planning for each unit's operation. Figure 2.1 lists some advantages when operators accurately predict the number of guests to be served in any future time period.

## Sales Histories

Maintaining an accurate sales history requires the systematic recording of all sales achieved during a predetermined time. It should be an accurate record of what items and the number of the items the operation has sold to its guests. Before operators can develop a sales history, however, they must consider the definition

---

1) Accurate revenue estimates
2) Improved ability to predict expenses
3) Greater efficiency in scheduling needed employees
4) Greater efficiency in scheduling menu item production schedules
5) Better accuracy in purchasing the correct amount of food for immediate use
6) Improved ability to maintain proper levels of perishable and nonperishable food inventories
7) Improved budgeting ability
8) Lower selling prices for guests because of increased operational efficiencies
9) Increased dollars available for current facility maintenance and future growth
10) Increased profit levels

---

**Figure 2.1**   Advantages of Accurate Sales Forecasts

of sales most helpful to them and ensure an understanding of exactly how their operation functions.

The simplest type of sales history records revenue only. The sales history format used by The Kebab Stop, a food truck, is shown in Figure 2.2. It is typical of the format used by an operation recording its sales revenue on a daily and weekly basis.

Note that the operator of the Kebab Stop might determine daily sales either from the POS system or from manually adding the information recorded on paper guest checks. They can then transfer that number on a daily basis to their sales history by entering the amount of their daily sales in the column titled Daily Sales.

**Sales to date** is the cumulative total of sales reported in the unit. To calculate this, the operator adds today's Daily Sales to the sales of all prior days in the **accounting period:** the time period for which sales records are being maintained.

To illustrate, review Sales to Date on Tuesday, January 2 in Figure 2.2. It is computed, for example, by adding Tuesday's sales to those of Monday (the prior day) to arrive at a Sales to Date total of $1,826.27 ($851.90 Monday sales + $974.37 Tuesday sales = $1,826.27 Sales to Date). In effect, the Sales to Date column is a running total of the sales achieved by the food truck for the week.

**Key Term**

**Sales to date:** The cumulative sales figures reported during a defined financial accounting period.

**Key Term**

**Accounting period:** A period of time (e.g., an hour, day, week, or month) in which an operator wishes to analyze the revenue and expenses of a business.

| Sales Period | Date | Daily Sales | Sales to Date |
|---|---|---|---|
| Monday | 1-1 | $ 851.90 | $ 851.90 |
| Tuesday | 1-2 | 974.37 | 1,826.27 |
| Wednesday | 1-3 | 1,004.22 | 2,830.49 |
| Thursday | 1-4 | 976.01 | 3,806.50 |
| Friday | 1-5 | 856.54 | 4,663.04 |
| Saturday | 1-6 | 1,428.22 | 6,091.26 |
| Sunday | 1-7 | 1,241.70 | 7,332.96 |
| Week's Total | | | $7,332.96 |

**Figure 2.2** Kebab Stop Sales History

Should the food truck operator prefer it, the reporting period could be defined in time blocks other than one week. Common alternatives are meal periods (breakfast, lunch, dinner, and so forth), days, weeks, two-week periods, **28-day accounting periods**, months, quarters (three-month periods), or any other unit of time that helps the operator to better understand the business. Modern POS systems allow operators to choose the specific revenue reporting periods of most interest to them.

**Key Term**

**28-day accounting period:** An accounting period that is four weeks (28 days) in length instead of a calendar month that has between 28 and 31 days. There are 13 four-week periods instead of 12 monthly periods in one year when using this system.

In some non-profit and even some for-profit foodservice operations, operators will not have the ability to consider their sales in terms of revenue generated. Figure 2.3 is an example of the type of sales history operators might use when no cash sales are typically reported.

In this example, the operator of the Camp Victory two-week tennis camp is interested in recording sales based on serving periods rather than an alternative time frame such as a 24-hour (one-day) period. This approach is often used in settings such as all-inclusive hotels and resorts, extended care facilities for senior citizens, nursing homes, college residence halls, correctional facilities, military bases, hospitals, and summer camps where knowledge of the number of actual guests served during a given period is critical for planning purposes.

Given the data in Figure 2.3, the implications for the Tuesday staffing of food servers at the camp are evident. For example, fewer servers will likely be needed from 9:00 to 11:00 a.m. than from 7:00 to 9:00 a.m. for an obvious reason. On Monday, fewer campers (40) ate between 9:00 and 11:00 a.m. than between 7:00 and 9:00 a.m. (121). A knowledgeable operator managing this camp could either reduce serving staff during the slower service period or schedule these workers to perform some other necessary tasks.

Similarly, the operator might decide not to produce as many menu items for consumption during the 9:00 to 11:00 a.m. period. Doing so can make more efficient use of the operation's labor and food products. It is, then, simply easier to manage well when operators can answer the question "How many guests will I serve?"

Sales histories can be created to record revenue, guests served, or both. In all cases, however, it is important that operators keep good records of how much they have sold in the past. The reason: Doing so is one key to accurately predicting the amount of sales the operation will likely achieve in the future.

The two major types of sales history averages utilized by foodservice operators are:

✓ Fixed average
✓ Rolling average

| | Campers Served | | | | | | | |
|---|---|---|---|---|---|---|---|---|
| Serving Period | Mon | Tues | Wed | Thurs | Fri | Sat | Sun | Total |
| 7:00–9:00 A.M. | 121 | | | | | | | |
| 9:00–11:00 A.M. | 40 | | | | | | | |
| 11:00–1:00 P.M. | 131 | | | | | | | |
| 1:00–3:00 P.M. | 11 | | | | | | | |
| 3:00–5:00 P.M. | 42 | | | | | | | |
| 5:00–7:00 P.M. | 161 | | | | | | | |
| Total Served | 506 | | | | | | | |

**Figure 2.3**   Camp Victory Sales History

## Fixed Averages

An **average**, also called a "mean," is a value computed by identifying the number of items in a series (example: 7 operating days in the week), adding the quantities of each item in the series (example: adding sales totals for each day), and then dividing the total of the quantities (sales totals) by the number of items in the series (example: if total sales for the 7 days is $1925 then $1925/7 = $275.00).

A **fixed average** is an average for a specific (fixed) time. Example: the first 14 days of a given month. When using this type of fixed average an operator computes the average amount of sales or guest activity for this 14-day period.

Note that this average is called "fixed" because the first 14 days of the month will always consist of the same days (the 1st through the 14th). Figure 2.4 illustrates this approach by detailing the sales activity of Benji's Bagels Shop. The cal-culation of this average (total revenue/number of days) is fixed, or constant, because Benji's operators have identified the 14 specific days used to make up the average.

**Key Term**

**Average (mean):** The value arrived at by adding the quantities in a series and dividing the sum of the quantities by the number of items in the series. Also commonly referred to as a "mean."

**Key Term**

**Fixed average:** The average amount of sales or volume over a specific series or time. For example: the first month of the year or the first week of the month.

If the operators wish, the fixed average number ($427.14; see the last line in Figure 2.4) could be used as a good predictor of the average revenue amount to be expected for each of the first 14 days of the next upcoming month.

## Rolling Averages

A **rolling average** is the average amount of sales or volume over a changing time period.

A fixed average is computed using a specific or constant set of data, but a rolling average is computed using data that will continually change. In most cases those operators who prefer to use a rolling average do so because this type of average helps smooth out data by constantly creating an updated average sales amount. By calculating the moving average, the impact of random short-term (one or two day) fluctuations in sales are reduced, and more weight is given to current, rather than older, data.

## Key Term

**Rolling average:** The average amount of sales or guest volume over a changing time period (for example, the last 10 days or the last three weeks). Also referred to as a "moving average."

To illustrate the calculation of a rolling average, consider the case of Val Salazar, who operates a sports bar. Val is interested in knowing what the average revenue dollars were in her operation for each *prior* seven-day period. In this case, the prior seven-day period changes, or rolls forward by one day, as each day passes. Val could have been interested in her average daily revenue last week (fixed average), but she prefers to know her average sales for the last seven days. This means that she will, at times, be using data from both last week and this week to compute her operation's last seven-day rolling average.

To initiate and maintain her rolling average Val uses three simple steps:

*Step 1: Identify the number of days to be included in the rolling average.*

In this example Val has selected seven days to be included in the rolling average.

*Step 2: Calculate the average sales volume for the first seven days.*

| Day | Daily Sales |
|---|---|
| 1 | $ 350.00 |
| 2 | 322.00 |
| 3 | 388.00 |
| 4 | 441.00 |
| 5 | 419.00 |
| 6 | 458.00 |
| 7 | 452.00 |
| 8 | 458.00 |
| 9 | 410.00 |
| 10 | 434.00 |
| 11 | 476.00 |
| 12 | 460.00 |
| 13 | 418.00 |
| 14 | 494.00 |
| 14-Day Total | $5,980.00 |

$5,980 sales/14 days =
$427.14 sales per day (average)

**Figure 2.4** Benji's Bagels Shop 14-Day Fixed Average

Using the sales data recorded in Figure 2.5, the first seven-day average for Val's Sports Bar is $4040, as shown in Figure 2.6.

*Step 3: Update the rolling average each day.*

On each successive day, Val will drop the oldest day's sale's total, and add the newest day's sales data.

For example, as shown in Figure 2.6, Val's days 2 thru 8 rolling average is $4,197.14.

Val can maintain an accurate rolling average simply by repeating Step Three described above.

The rolling average, although more complex and time-consuming to calculate than a fixed average, can be extremely useful in recording data to help operators make effective predictions about the future sales levels they can expect. This is true because, in many cases, rolling data are more current and therefore more relevant than some fixed historical averages. In most cases foodservice operators will find it advantageous and time saving to use an electronic spreadsheet program such as

| Day | Sales |
|---|---|
| 1 | $3,500.00 |
| 2 | $3,200.00 |
| 3 | $3,900.00 |
| 4 | $4,400.00 |
| 5 | $4,200.00 |
| 6 | $4,580.00 |
| 7 | $4,500.00 |
| 8 | $4,600.00 |
| 9 | $4,100.00 |
| 10 | $4,400.00 |
| 11 | $4,700.00 |
| 12 | $4,600.00 |
| 13 | $4,180.00 |
| 14 | $4,940.00 |

**Figure 2.5** Val's Sports Bar 14-Day Sales Levels

| Day | 1–7 | 2–8 | 3–9 | 4–10 | 5–11 | 6–12 | 7–13 | 8–14 |
|---|---|---|---|---|---|---|---|---|
| 1 | $ 3,500 | – | | | | | | |
| 2 | 3,200 | $ 3,200 | – | | | | | |
| 3 | 3,900 | 3,900 | $ 3,900 | – | | | | |
| 4 | 4,400 | 4,400 | 4,400 | $ 4,400 | – | | | |
| 5 | 4,200 | 4,200 | 4,200 | 4,200 | $ 4,200 | – | | |
| 6 | 4,580 | 4,580 | 4,580 | 4,580 | 4,580 | $ 4,580 | – | |
| 7 | 4,500 | 4,500 | 4,500 | 4,500 | 4,500 | 4,500 | $ 4,500 | – |
| 8 | | 4,600 | 4,600 | 4,600 | 4,600 | 4,600 | 4,600 | $ 4,600 |
| 9 | | | 4,100 | 4,100 | 4,100 | 4,100 | 4,100 | 4,100 |
| 10 | | | | 4,400 | 4,400 | 4,400 | 4,400 | 4,400 |
| 11 | | | | | 4,700 | 4,700 | 4,700 | 4,700 |
| 12 | | | | | | 4,600 | 4,600 | 4,600 |
| 13 | | | | | | | 4,180 | 4,180 |
| 14 | | | | | | | | 4,940 |
| Total | $ 28,280 | $ 29,380 | $ 30,280 | $ 30,780 | $ 31,080 | $ 31,480 | $ 31,080 | $ 31,520 |
| 7-Day | $4,040.00 | $4,197.14 | $4,325.71 | $4,397.14 | $4,440.00 | $4,497.14 | $4,440.00 | $4,502.86 |

**Figure 2.6** Val's Sports Bar Seven-Day Rolling Average

Microsoft Excel™ or Apple Numbers™ to simplify the process of maintaining the rolling average.

To best predict future sales operators may choose to compute fixed averages for some time periods and rolling averages for others. For example, it may be helpful for an operator to know their average daily sales for the first 14 days of last month and their average sales for the most recently past 14 days. If, for example, these two numbers are very different, the operator will know whether the number of sales they can expect in the future is increasing or declining. Regardless of the type of average operators decide to use, they should always document their sales history because it is from this data that they will be better able to predict their future sales levels.

## Average Sale Per Guest

Some foodservice operations do not record revenue as the primary measure of their sales activity. For them, developing sales histories by recording the number of individuals they serve each day makes the most sense. Therefore, **guest count,** the term used in the foodservice industry to indicate the number of people served, is recorded on a regular basis.

Guest count may be particularly important in operations that do not collect revenue at the time of sale.

Some foodservice operators may decide their businesses are best managed by tracking both revenue generated and counts of the number of guests. If they do, they will record both revenue and guest counts, and then they will have the information that is needed to compute their **average sale per guest**.

The formula used to calculate average sale per guest is:

**Key Term**

**Guest count:** The number of individuals served in a defined accounting period.

**Key Term**

**Average sale per guest:** The mean amount of money spent per guest during a defined accounting period. Also commonly referred to as "check average."

$$\frac{\text{Total Sales}}{\text{Number of Guests Served}} = \text{Average Sale per Guest}$$

The average sale per guest (check average) is one of a foodservice operator's most commonly used formulas. To compute their operation's average sale per guest, managers must do three things:

Step 1. Identify the amount of sales generated during a specific time period.

Step 2. Identify the number of guests served in the same time period.

Step 3. Apply the Average Sale per Guest formula.

For example, assume an operator determined the amount of sales generated last week was $18,750 (Step 1). Further, assume that 1,500 guests were served last week (Step 2). Applying the Average Sale per Guest formula (Step 3) the operator would find that last week's average sale per guest was $12.50.

$$\frac{\$18,750 \text{ Total sales last week}}{1,500 \text{ Guests served last week}} = \$12.50 \text{ average sale per guest}$$

One reason why the ability to accurately calculate an operation's average sale per guest is important is that foodservice operators are often evaluated on their efforts to maintain or increase the average amount spent by each guest visiting the operations.

To illustrate the importance of identifying an operation's average sale per guest, consider the information shown in Figure 2.7, in which the operator of Agueda's Portuguese Restaurant decided to monitor and record:

1) Sales
2) Guests served
3) Average sale per guest

Most POS systems are programmed to report the amount of revenue generated during a selected time, the number of guests served, and the average sale per guest. In the case of Agueda's Portuguese Restaurant, Monday's revenue was $1,365; it served 190 guests, and the average sale per guest that day was $7.18 ($1,365 sales/190 guests served = $7.18). On Tuesday, the average sale per guest was $8.76 ($2,750 sales/314 guests served = $8.76).

To compute the two-day revenue average, the operator at Agueda's would add Monday's revenue and Tuesday's revenue and then divide by 2, yielding a two-day revenue average of $2,057.50 [($1,365 + $2,750)/2 = $2,057.50].

| Sales Period | Date | Day | Sales | Guests Served | Average Sale per Guest |
|---|---|---|---|---|---|
| Monday | Jan 1 | Monday | $1,365.00 | 190 | $7.18 |
| Tuesday | Jan 2 | Tuesday | 2,750.00 | 314 | 8.76 |
| Two-Day Average | | | 2,057.50 | 252 | 8.16 |

**Figure 2.7** Agueda's Portuguese Restaurant Sales History

In a like manner, the two-day average for the number of guests served is computed by adding the number of guests served on Monday to the number served on Tuesday and then dividing by 2, yielding a two-day average of guests served of 252 [(190 + 314)/2 = 252].

It might be logical to think that the operator of Agueda's could simply compute the Monday and Tuesday combined average sale per guest by adding the averages from each day and then dividing by 2. It is important to understand that this would *not* be correct.

A formula consisting of Monday's average sale per guest plus Tuesday's average sale per guest divided by 2 [($7.18 + $8.76)/2] yields $7.97. In fact, the actual two-day average sale per guest is $8.16 [($1,365 + $2,750)/(190 + 314) = ($4,115/504) = $8.16].

Although the difference of $0.19 might, at first glance, appear to be inconsequential, assume you are the president of a restaurant chain with 5,000 units worldwide. If each unit served 1,000 guests per day and you miscalculated the average sale per guest by $0.19, your daily revenue calculation would be "off" by $950,000 per day [(5,000 units × 1,000 guests × $0.19) = $950,000]!

Returning to the Agueda's Portuguese Restaurant example, the correct procedure for computing the two-day average sale per guest would be:

$$\frac{(\text{Monday Sales} + \text{Tuesday Sales})}{(\text{Monday Guests} + \text{Tuesday Guests})} = \text{Two-Day Average Sale per Guest}$$

Or

$$\frac{(\$1,365 + \$2,750)}{(190 + 314)} = \$8.16$$

The correct computation of average sale per guest in this example is a **weighted average**.

To further demonstrate the importance of utilizing weighted averages, consider the data in Figure 2.8 and assume that an operator wanted to answer the question, "What is the combined average sale per guest for these two days?"

From the data in Figure 2.8, it is easy to see that the two-day average would *not* be $7.50 [($5.00 + $10.00)/2 = $7.50] because many more guests were served on the day the average sale per guest was $10.00 than on the day the average sale per guest was $5.00. With so many guests spending an average of $10.00, and so few spending an average of $5.00, the overall average should be much closer to $10.00 than to $5.00.

**Key Term**

**Weighted average:** An average that combines data on the number of guests served and how much each has spent during a specific accounting period.

|  | Sales | Guests Served | Average Sale per Guest |
|---|---|---|---|
| Day 1 | $ 100 | 20 | $ 5.00 |
| Day 2 | $4,000 | 400 | $10.00 |
| Two-Day Average | $2,050 | 210 | ??? |

**Figure 2.8** Weighted Average Calculation

In fact, utilizing the average sale per guest formula, the correct weighted average sale per guest would be $9.76, as follows:

$$\frac{(\text{Day 1 Sales} + \text{Day 2 Sales})}{(\text{Day 1 Guests} + \text{Day 2 Guests})} = \text{Two-Day Average Sale per Guest}$$

Or

$$\$100 + \$4,000/20 + 400 \text{ guests} = \$9.76 \text{ per guest}$$

Although a sales history may consist of revenue, number of guests served, and/or average sale per guest, depending upon the type of facility being managed, operators may want to know even more detailed information about their sales. This may include information such as the number of guests served in a specific meal or time period (e.g., breakfast, lunch, or dinner), the method of meal delivery (e.g., drive-thru sales vs. carry out or dine-in sales), or the method used to order (online vs. on-site).

The important concept to remember is that all operators should develop the sales history information that best suits their own business. That information should be updated at least daily, and a cumulative total for the appropriate accounting periods should also be maintained. In most cases, sales histories should be kept for a period of at least two years. This allows an operator to have a good sense of what has happened in their business in the recent past.

Of course, if an operator manages a newly opened operation, or one that has recently undergone a major concept change, they will not have the advantage of reviewing meaningful sales histories because these histories will not exist. In that situation, operators must begin to build and maintain their sales histories as soon as possible. Doing so will enable them to have useful sales information on which to base future managerial decisions as quickly as possible.

| Technology at Work |
| --- |
| Modern POS systems can provide foodservice operators with a variety sales history reports. |
| These include reports on daily sales and sales totals by week and month. They also can report sales by dining room section or even specific tables. They can be used to report sales by menu category (e.g., entrees, soups, salads, and desserts). |
| In addition, sales reports can be created specifically for dine-in, take out, and delivery services, allowing operators to carefully examine exactly from where their business comes. |
| Some POS systems allow operators to track sales by server. This allows operators to know about the relative sale per guest totals achieved by their various waitstaff members |
| To learn more about these and other sales reporting features of POS systems enter "POS system sales reports for restaurants" in your favorite search engine and review the results. |

## Sales Variance

After an accurate sales history recording process is established, most operators begin to see that their businesses experience some level of **sales variance**. These sales variances are normal and give operators an indication of whether their sales are increasing, declining, or staying the same. Since variance-related information is so important to pre-

**Key Term**

**Sales variance:** An increase or decrease from previously experienced or predicted sales levels.

dicting future sales levels, many operators enhance their sales history information by including sales variance as an additional component of the history.

Figure 2.9 details a portion of a sales history that has been modified to include a sales "Variance" column, and this allows the business operator to see how sales differ from a prior period.

| Month | Sales This Year | Sales Last Year | Variance |
| --- | --- | --- | --- |
| January | $ 54,000 | $ 51,200 | $ 2,800 |
| February | 57,500 | 50,750 | 6,750 |
| March | 61,200 | 57,500 | 3,700 |
| First-Quarter Total | 172,700 | 159,450 | 13,250 |

**Figure 2.9**   Chinese Garden's Sales History and Variance

In this example, the operator of the Chinese Gardens restaurant wants to compare sales for the first three months of this year to sales for the first three months of last year, as recorded in that period's sales history report.

The variance in Figure 2.9 is determined by subtracting sales last year from sales this year. In January, the variance figure is calculated as:

Sales This Year − Sales Last Year = Variance

Or

$54,000 − $51,200 = $2,800

The operator of the Chinese Garden can see that the sales for the first quarter are greater than last year's first quarter. In fact, all three months in the first quarter of the year showed revenue increases over the prior year.

The total sales improvement for the first quarter was $13,250 ($172,700 − $159,450 = $13,250). The sales history and variance format used in Figure 2.9 lets an operator know the dollar value of revenue variance. However, many good operators will want to know even more because simply knowing the dollar value of a variance has limitations.

To illustrate, consider two restaurant operators. One operator's restaurant had revenue of $1,000,000 last year. The second operator's restaurant generated one-half as much revenue, or $500,000. This year both operators had sales increases of $50,000.

However, it is clear that, while both experienced a $50,000 sales increase, that increase represents a much greater proportional change in the second operation than in the first. Effective operators are often interested in the **percentage variance,** or percentage change, in their sales during one time period when compared to a different time period.

Figure 2.10 shows how the sales history data at the Chinese Garden can be expanded to include percentage variance as part of that operation's complete sales history.

**Key Term**

**Percentage variance:** The change in sales (or expense) expressed as a percentage, and which results from comparing two different operating periods.

| Month | Sales This Year | Sales Last Year | Variance | Percentage Variance |
|---|---|---|---|---|
| January | $ 54,000 | $ 51,200 | $ 2,800 | 5.5% |
| February | 57,500 | 50,750 | 6,750 | 13.3 |
| March | 61,200 | 57,500 | 3,700 | 6.4 |
| First-Quarter Total | 172,700 | 159,450 | 13,250 | 8.3 |

**Figure 2.10** Chinese Garden Sales History, Variance, and Percentage Variance

Percentage variance is obtained by subtracting sales last year from sales this year and then dividing the resulting number by last year's sales. For example, in January, the percentage variance is calculated as:

$$\text{Sales This Year} - \text{Sales Last Year/Sales Last Year} = \text{Percentage Variance}$$

Or

$$\frac{\$54,000 - \$51,200}{\$51,200} = 0.055\,(5.5\%)$$

Note that the resulting decimal form percentage can be converted to the more frequently used common form by moving the decimal point two places to the right, or when it is multiplied by 100.

Returning to the previous example of the two restaurant operators, each of whom achieved \$50,000 revenue increases, the percentage variance formula makes it clear that the operation with higher sales increased its revenue by 5.0 percent [(\$1,050,000 – \$1,000,000)/\$1,000,000 = .05 (5%). In contrast, the operation with less original revenue and the same \$50,000 increase in sales achieved a 10% increase [(\$550,000 – 500,000)/\$500,000 = .10] or 10%. As an operator's expertise increases, they will typically find additional areas in which knowing the percentage variance in revenue and/or an expense (addressed later in this book) will greatly assist in their cost control-related decision making.

---

**Technology at Work**

In nearly all cases the sales history that will be of most use to a foodservice operator will be created and maintained internally. While an operator's point of sale (POS) systems can provide a great deal of raw sales data, the needs of an individual operator will typically require them to create a sales history spreadsheet unique to their own situation.

While there are numerous excellent spreadsheet programs available for use, one of the most popular is the Microsoft Excel program. Excel is available in both Mac and PC formats and learning how to use Excel to build spreadsheets is a skill that will be valuable in other cost control-related efforts.

Fortunately, there are instructional resources available to operators who want to learn how to use Excel, and YouTube is a good source for them. To become familiar with the basics of utilizing Excel, go to YouTube. When you arrive, enter "Excel for beginners" in the search bar and review one or more instructional videos.

# Sales Forecasts

Outstanding foodservice operators can (almost!) see the future as they plan for the sales they will achieve and the number of guests they expect to serve. Every operator can learn to do this, however, when they apply information from their sales histories and a knowledge of percentage variance to estimating their operation's future sales.

### Forecasting Future Sales

Depending on the type of facility they manage, operators may be interested in predicting, or forecasting, future sales (revenue), guest counts, and/or average sale per guest. To illustrate how this is done, consider Lyla Metzger, the manager of the Meeting Grounds coffee shop adjacent to the campus of State College.

Lyla's guests consist of college students, most of whom come to the Meeting Grounds to talk, surf the web, text, eat, and study. Lyla has done a good job maintaining sales histories in the two years she has managed the business. She daily records the revenue dollars and the number of students frequenting the operation. Revenue data for the last three months of the year are shown in Figure 2.11.

As seen in Figure 2.11, fourth-quarter revenues for Lyla's operation have increased from the previous year, and there might be a variety of reasons for this. For example, Lyla may have extended her hours of operation to attract more students. She may have increased the size of coffee beverages but did not increase her prices to create greater value for her guests. As a final example, perhaps a competing coffee shop closed during this time period.

Using knowledge of her own operation and the market, Lyla wants to predict the sales level she will likely experience in the first three months of next year. This sales forecast will be most helpful as she plans for next year's anticipated expenses, optimized staffing levels, and anticipated profits.

| Month | Sales This Year | Sales Last Year | Variance | Percentage Variance |
|---|---|---|---|---|
| October | $ 75,000 | $ 72,500 | $ 2,500 | 3.4% |
| November | 64,250 | 60,000 | 4,250 | 7.1 |
| December | 57,500 | 50,500 | 7,000 | 13.9 |
| Fourth-Quarter Total | 196,750 | 183,000 | 13,750 | 7.5 |

**Figure 2.11**  Meeting Grounds Revenue History

| Month | Sales Last Year | % Increase Estimate | Increase Amount | Revenue Forecast |
|---|---|---|---|---|
| January | $ 68,500 | 7.5% | $ 5,137.50 | $ 73,637.50 |
| February | 72,000 | 7.5 | 5,400.00 | 77,400.00 |
| March | 77,000 | 7.5 | 5,775.00 | 82,775.00 |
| First-Quarter Total | 217,500 | 7.5 | 16,312.50 | 233,812.50 |

**Figure 2.12**   Meeting Grounds First-Quarter Revenue Forecast

The first question Lyla must address is the amount her sales increased. Revenue increases range from a low in October of 3.4 percent, to a high in December of 13.9 percent. The overall quarter (3-month) average of 7.5 percent is the figure that Lyla elects to use as she predicts her sales revenue for the first quarter of the coming year. She feels it is neither too conservative, as would be the case if she used the October percentage increase, nor too aggressive, as would be the case if she used the December figure.

If Lyla were to use the 7.5 percent average increase from the fourth quarter of last year to predict her revenues for the first quarter of this year, a revenue forecast could be developed for the first quarter of next year as shown in Figure 2.12.

The revenue forecast for this period is determined by multiplying sales last year by the percentage increase estimate, and then adding the percentage increase amount to sales last year.

For January, the revenue forecast is calculated using the following formula:

Sales Last Year + (Sales Last Year × % Increase Estimate) = Revenue Forecast

Or

$$\$68,500 + (\$68,500 \times 0.075) = \$73,637.50$$

An alternative way to compute the revenue forecast is to use a math shortcut as follows:

Sales Last Year × (1 + % Increase Estimate) = Revenue Forecast

Or

$$\$68,500 \times (1 + 0.075) = \$73,637.50$$

In this example, Lyla is using the increases the operation experienced in the past to predict increases she may experience in the future. Monthly revenue figures from last year's sales history plus percentage increase estimates based on those histories can give Lyla a good idea of the revenue levels she might achieve in January, February, and March of the coming year.

---

**Find Out More**

Foodservice operators desiring to forecast future sales must stay up-to-date on changing consumers' desires and tastes. To help them do that, the National Restaurant Association (NRA) produces an annual report called "The What's Hot Report" designed to help operators recognize changes in consumer tastes and product demand.

"Understanding changing consumer desires is essential to the success of restaurants in every community across the country," says Michelle Korsmo, President & CEO of the National Restaurant Association. "The What's Hot report provides an invaluable lens through which operators can evaluate and adapt emerging trends to create dining experiences that stimulate and engage their consumers, and perhaps even push the envelope forward on what's hot next year."[1]

More than 500 professional chefs from the American Culinary Federation (ACF) and NRA members with chef-related titles provide insights that support a comprehensive outlook of the leading food and menu trends.

To learn more about this useful reporting of dining trends, enter "NRA What's Hot For (insert year)" in your favorite search engine and review the results.

---

## Forecasting Future Guest Counts

Using the same techniques employed to estimate increases in sales, operators can also estimate increases or decreases in the number of guests they will serve. Figure 2.13 shows how Lyla, the manager of the Meeting Grounds, used last year's fourth quarter guest counts to determine the variance and percentage variance for the guest counts she achieved in her facility in the fourth quarter of this year.

If Lyla were to use the 6.1 percent average increase from the fourth quarter of last year to predict her guest counts for the first quarter of the coming year, a planning spreadsheet could be developed as presented in Figure 2.14. Note: Lyla is not required to use the same percentage increase estimate for each coming month. Any forecasted increase that management feels is appropriate can be used to predict future sales.

---

1 https://restaurant.org/research-and-media/media/press-releases/national-restaurant-association-releases-2023-whats-hot-survey/

| Month | Guests This Year | Guests Last Year | Variance | Percentage Variance |
|---|---|---|---|---|
| October | 14,200 | 13,700 | + 500 | 3.6% |
| November | 15,250 | 14,500 | + 750 | 5.2 |
| December | 16,900 | 15,500 | +1,400 | 9.0 |
| Fourth-Quarter Total | 46,350 | 43,700 | +2,650 | 6.1 |

**Figure 2.13**   Meeting Grounds Guest Count History

| Month | Guests Last Year | % Increase Estimate | Guest Increase Estimate | Guest Count Forecast |
|---|---|---|---|---|
| January | 12,620 | 6.1% | 770 | 13,390 |
| February | 13,120 | 6.1 | 800 | 13,920 |
| March | 13,241 | 6.1 | 808 | 14,049 |
| First-Quarter Total | 38,981 | 6.1 | 2,378 | 41,359 |

**Figure 2.14**   Meeting Grounds First-Quarter Guest Count Forecast

The guest count forecast is determined by multiplying last year's guest count by the percentage increase estimate this year, and then adding the result to the guest count last year. In the month of January, the guest count forecast is calculated using the following formula:

Guest Count Last Year + (Guest Count Last Year × %Increase Estimate) = Guest Count Forecast

Or

$$12,620 + (12,620 \times 0.061) = 13,390$$

This process can be simplified by using a math shortcut, as follows:

Guests Last Year × (1.00 + % Increase Estimate) = Guest Count Forecast

Or

$$12,620 \times (1.00 + 0.061) = 13,390$$

**What Would You Do? 2.1**

"A 90-minute wait! You've got to be kidding!" said the guest.

"I'm sorry sir," replied Rebecca, "We'll seat your party as quickly as possible."

Rebecca was the dining room supervisor at the Hummus Corner, a small restaurant that featured Middle Eastern and North African Cuisine. Guests loved the menu, and the property was often very busy. When that happened, the line of guests waiting to be seated got long and, sometimes, waiting customers got upset.

"Listen" replied the guest, "I understand when places are busy. It can take a while to serve everyone. But look, nearly half of your dining room is empty. The tables just need to be cleared and reset."

"Yes sir," replied Rebecca. "But our team is clearing and resetting the tables as fast as they can."

"Then you need more dining room help. We'll just come back another time," said the guest, as he left the restaurant along with his female dining companion, with two small children in hand.

"I'm really very sorry, sir," called out Rebecca as she watched him leave. Rebecca thought to herself, "This happens way too often!"

**Assume you were the owner of the Hummus Corner. How could accurate daily sales forecasts help you anticipate periods of high demand? What will be the likely long-term impact on the revenue-generating ability of your restaurant if you consistently understaff your dining rooms because your sales forecasts were either inaccurate or nonexistent?**

## Forecasting Future Average Sale Per Guest

The average sale per guest (check average) is the average amount of money each guest spends during a visit. The same formula is used to forecast an operation's future average sales per guest as was used in forecasting total revenue and guest counts. Using data taken from their sales histories, operators can forecast their future average sale per guest using the formula:

Last Year's Average Sale per Guest + Estimated Increase in Sale per Guest = Average Sale per Guest Forecast

Alternatively, an operator such as Lyla could compute the forecasted average sale per guest at Meeting Grounds by using the data collected from her revenue forecasts (Figure 2.12) and combining that with the guest count forecasts she created (Figure 2.14). If that is done, the average sale per guest forecast produced is shown in Figure 2.15.

In this example, the average sale per guest forecast is obtained by dividing an operation's revenue forecast by its guest count forecast. For example, the average sale per guest forecast for January for the Meeting Grounds would be calculated as:

| Month | Revenue Forecast | Guest Count Forecast | Average Sale per Guest Forecast |
|---|---|---|---|
| January | $ 73,637.50 | 13,390 | $5.50 |
| February | 77,400.00 | 13,920 | 5.56 |
| March | 82,775.00 | 14,049 | 5.89 |
| First-Quarter Total | 233,812.50 | 41,359 | $5.65 |

**Figure 2.15**  Meeting Grounds First-Quarter Average Sale per Guest Forecast

Revenue Forecast /Guest Count Forecast = Average Sale per Guest Forecast

Or

$$\frac{\$73,637.50 \text{ Revenue forecast}}{13,390 \text{ Guest forecast}} = \$5.50 \text{ per Guest}$$

Increasingly, sophisticated POS systems analyze not only how many guests were served on a given day in the past, but they also provide data that helps operators better understand why that number was served. For example, historical weather patterns can be analyzed to discover if rainy days affect total sales volume (and therefore might indicate that a forecast adjustment should be made on a day when rain is predicted). Weather, holidays, day of the month, and day of the week are just some of many factors that may affect sales volume and should be carefully considered by sophisticated foodservice operators as they prepare sales forecasts.

Operators should recognize that sales histories, regardless of how well they are developed and maintained, may not provide all the information needed to accurately predict future sales. An operator's knowledge of potential price changes, new or lost competitors, facility renovations, and improved selling programs are additional factors that operators might consider when they forecast future sales.

There is no question, however, that accurate sales history reports can be easily developed and will serve as the cornerstone of other financial control systems in a foodservice operation. Without accurate future sales forecasts, the control systems an operator implements, regardless of their sophistication, will perform poorly.

Properly maintained sales histories can help operators answer two of their most important cost control-related questions, namely, "How many people are coming tomorrow?" and "How much is each person likely to spend?" The judgment of an operator is critical in forecasting answers to these questions.

Foodservice operators must develop an efficient and cost-effective way of serving their guests. They must be ready to provide arriving guests with quality food and beverage products and enough staff to serve them properly. Operators who have properly forecasted the number of individuals coming to their businesses and the amount of money these guests are likely to spend, must next purchase the ingredients needed to prepare the menu items their guests purchase. Doing so cost-effectively is the sole topic addressed in the next chapter!

---

**Find Out More**

Some foodservice operators find that the math required to properly manage their businesses can be challenging. The calculation of commonly used averages, percentages, and variance percentages can sometimes be puzzling until they have been mastered.

There are several publications that explain and illustrate the mathematical formulas typically used in a foodservice operation. One of the best is the book *Culinary Math*.

Written by two former instructors at The Culinary Institute of America (CIA), this book is an indispensable math resource for foodservice professionals. Covering topics such as calculating yield percent, determining portion costs, changing recipe yields, and converting between metric and U.S. measures, it offers a review of math basics, easy-to-follow lessons, and detailed examples. The book also has newly revised practice problems in every chapter.

To review the contents of this useful book, go to "www.wiley.com." When you arrive at the Wiley website, enter "Culinary Math" in the search bar to examine the outline and content of this valuable industry resource.

---

**What Would You Do? 2.2**

"This could be a real problem," said James.

"What is it?" asked Tara.

It was 6:00 p.m. on Tuesday and James, the kitchen manager, had just answered the phone at Larson's Diner. Larson's was a 100-seat old fashioned diner that featured traditional American diner food along with their specialty Ice Cream Sundaes.

"Well, I just got a call from the Trail Lines Bus company," said James. "They have a full 56-person tour bus and driver, and the driver just told me that they will be here in 20 minutes. Seems they saw our website, the group liked it, and they want to stop for dinner."

"Wow," said Tara, the diner's manager "that's a bunch of people! Tuesdays are normally pretty slow, and we only have three servers and two cooks on tonight. You're right! This could be a real problem."

"Well," said James, "we have a about 20 minutes to figure it out!"

**Assume you were Tara. What steps could you take to help ensure quality service at your operation when your actual sales levels are much higher (or much lower) than your original sales forecast? What could be additional examples of unforeseen or uncontrollable factors that could significantly affect the accuracy of your daily sales forecast?**

## Key Terms

| | | |
|---|---|---|
| Sales history | 28-day accounting period | Average sale per guest |
| Sales forecast | Average (mean) | Weighted average |
| Sales volume | Fixed average | Sales variance |
| Sales to date | Rolling average | Percentage variance |
| Accounting period | Guest count | |

### Operator's 10-Point Tactics for Success Checklist

Evaluate your need for, and the current status of, each of the following operational tactics. For those tactics you think are important, but not yet in place, develop an action plan for its implementation including who will be responsible for the tactic's completion and the target date by which it should be completed.

| | | | | If Not Done | |
|---|---|---|---|---|---|
| Tactic | Don't Agree (Not Done) | Agree (Done) | Agree (Not Done) | Who Is Responsible? | Target Completion Date |
| 1) Operator recognizes the importance of accurate sales forecasts to the effective control of operating costs. | ---- | ---- | ---- | | |
| 2) Operator can identify specific advantages of creating accurate sales forecasts. | ---- | ---- | ---- | | |
| 3) Operator can apply the formula utilized to calculate a fixed average. | ---- | ---- | ---- | | |
| 4) Operator can apply the formula utilized to calculate a rolling average. | ---- | ---- | ---- | | |
| 5) Operator knows and can apply the formula utilized to calculate average sale per guest. | ---- | ---- | ---- | | |

| Tactic | Don't Agree (Not Done) | Agree (Done) | Agree (Not Done) | If Not Done | |
|---|---|---|---|---|---|
| | | | | Who Is Responsible? | Target Completion Date |
| 6) Given the required historical data operator can calculate the dollar amount of a sales variance. | ____ | ____ | ____ | | |
| 7) Given the required historical data, the operator can calculate a percentage variance in sales. | ____ | ____ | ____ | | |
| 8) Operator can create an accurate future sales forecast. | ____ | ____ | ____ | | |
| 9) Operator can create an accurate future guest count forecast. | ____ | ____ | ____ | | |
| 10) Operator can create an accurate future average sale per guest forecast. | ____ | ____ | ____ | | |

# 3

## Cost Control in Purchasing

---

**What You Will Learn**

1) The Importance of Professional Purchasing
2) How to Determine What to Buy
3) How to Determine How Much to Buy
4) How to Use Purchase Orders to Create Purchasing Records

---

**Operator's Brief**

In this chapter, you will learn that proper purchasing is one of the most important skills you can develop. Knowing the objectives of a professional purchasing program and the steps required in the process will allow you to select and purchase the right products, at the right price, and have them delivered to you when they are needed.

The initial step in a professional purchasing program will require you to identify the products you need to produce the menu items you sell. This is done by establishing a purchasing specification (spec) for each product you will buy. The development of product specs is necessary to ensure that suppliers deliver only those items that you have specifically determined are best for the menu items you will produce and serve to your guests.

After you have determined what you should buy, the next step in the process requires that you determine how much to buy. This is done by estimating future sales of the menu items you offer and by establishing proper inventory levels for items you hold in storage.

After determining what you should buy, and how much of it you should buy, placing your actual order is the next step in the process. Prior to placing orders, however, you must select suppliers and understand how your chosen suppliers can be partners in your cost control efforts.

The actual placement of orders is done through the development of purchase orders (POs). The use of a formal PO will provide you with a record of what you have ordered, how much has been ordered, and the prices you will pay.

---

**CHAPTER OUTLINE**

---

The Importance of Controlling Costs in Purchasing
    Objectives of Professional Purchasing
    Steps in the Professional Purchasing Process
Determining Needed Products
Determining Needed Product Quantities
    Forecasting Menu Item Sales
    Establishing Inventory Levels
Placing the Purchase Order
    Choosing Suppliers
    Understanding Suppliers
    Recording Purchases

---

# The Importance of Controlling Costs in Purchasing

For all foodservice operators, controlling product and production costs begins with effective **purchasing**. Purchasing relates to an operator's "buying" activities including determining the products and amounts to buy, placing orders, and paying suppliers for the delivered items.

**Key Term**

**Purchasing:** The process of "buying": placing an order, receiving a product (or service), and paying the supplier.

## Objectives of Professional Purchasing

The best foodservice operators recognize that professional purchasing is a process that must meet important objectives. Figure 3.1 summarizes the objectives of a cost-effective purchasing program:

Foodservice operators have one primary purchasing concern: to identify and obtain the products that best allow them to meet the wants and needs of their guests at a price that provides good value to the buyer.

The purchasing process is never-ending. Guest preferences change and new product and service alternatives are continually introduced. Price concerns require ongoing attention and, as a result, foodservice operators must continually learn about the marketplace and its suppliers, revise their purchasing procedures as needed, and evaluate their success.

## Steps in the Professional Purchasing Process

The effective control of purchasing in a foodservice operation is a multi-step process:

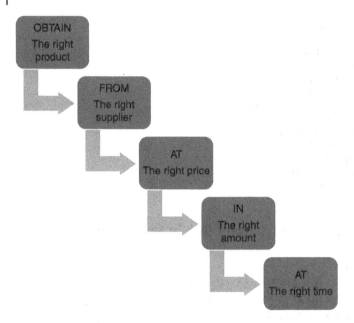

**Figure 3.1** Objectives of Professional Purchasing

### Step 1: Identify Product Needs

Product need is driven primarily by what an operation's guests want to buy. Ingredients needed to make popular menu items must be readily available to production staff in the right amounts and at the right time.

### Step 2: Determine Quality Requirements

While the term **quality** is used frequently in the foodservice industry, from a purchasing perspective, quality refers to the appropriateness of an item for its intended use. Quality considers how suitable a product or service is for its intended purpose. For example, ground beef can be purchased with different percentages of fat content, and it can be purchased fresh or frozen.

**Key Term**

**Quality:** Suitability for intended use. The closer an item comes to being suitable for its intended use, the more appropriate is its quality.

In this example, a foodservice operator must choose the quality of ground beef appropriate for the items to be made from that ingredient. Sometimes, the quality of an item may be suitable use in one recipe but unsuitable for use in another recipe.

### Step 3: Determine Quantities to Purchase

In many cases, products used in a foodservice operation can be purchased in large or small amounts. In most cases, buying in larger amounts reduces per unit buying costs. However, this may also result in excessive product spoilage if the item cannot be held for long periods of time without a loss of quality. Choosing the appropriate quantity to purchase is directly affected by several factors including storage availability, delivery frequency, and an operator's forecast of future sales.

### Step 4: Identify Supplier Sources

In many cases, a foodservice operator will have more than one possible supplier for specific items. An operator's experience, supplier references, and "trial orders" with potential suppliers can help to determine a "short list" of those from whom orders should be placed.

### Step 5: Select Supplier for Orders

In this step, operators choose an appropriate supplier. The answer to which supplier should be chosen depends, in part, upon what is to be purchased. For example, an operator may use one supplier for meats and fish, and another supplier for fresh produce items. It is also possible that more than one supplier carrying needed items can provide them, and a purchase decision focusing on best price is selected for the current supply of items that are needed.

### Step 6: Order Products

Products can be ordered after the proper quality and quantity are known (Steps 2 and 3) and after the supplier has been selected (Step 5). Today's foodservice operators may place their orders via a supplier's website, by telephone, and/or in person. In all cases, operators should make a record of the items they are buying. This is done with a formal **purchase order (PO)**. Note: POs will be addressed in detail later in this chapter.

**Key Term**

**Purchase order (PO):** A detailed listing of products a buyer is purchasing from a supplier.

### Step 7: Receive Products

Requested products must be delivered, and proper receiving procedures (see Chapter 4) are needed to ensure that the proper quality and quantity of products are received. Then, products must be properly stored to minimize quality or theft problems. Appropriate recordkeeping tasks applicable to product storage are needed in this step.

### Step 8: Pay for Products

The timing of payments, concerns about fraud as payments are processed, and accounting concerns to identify and assign costs to the specific operating departments that incurred them become important when payments are made.

### Step 9: Evaluate the Purchasing Process

Evaluation is useful to ensure that each step in the purchasing process is done correctly. This concern applies to the way that basic purchasing process decisions are made (Steps 1–3) and to activities for specific orders (Steps 4–9).

## Determining Needed Products

The identification of product needs in a foodservice operation begins with determining the menu items to be sold and the quality levels of products needed to produce the menu items. The question of what food or beverage product should be purchased is answered with a **product specification (spec)** for each needed item.

A product spec allows foodservice operators to precisely communicate their product needs with suppliers so they can receive and prepare the exact item every time it is ordered.

**Key Term**

**Product specification (spec):** A detailed description of a recipe ingredient or complete menu item to be served.

A product spec determines neither the best product to buy nor the product that costs the least. Rather, it is the product determined to be the *most appropriate* for its intended use in terms of both quality and cost.

A product spec that is not sufficiently detailed can create problems because the needed level of item quality may be unknown and/or not delivered. However, if product specs are written too tight (they are overly and unnecessarily specific), then too few suppliers may be able to provide the needed products. The result: excessively high costs from the few suppliers who can deliver the items.

Figure 3.2 is an example of a professionally prepared product specification for the bacon used by a foodservice operation for bacon, lettuce, and tomato (BLT) sandwiches:

**Key Term**

**Product yield:** The servable amount of a raw ingredient remaining after it has been cleaned, trimmed, cooked, and portioned.

| Product Name: | Bacon, sliced | Spec #: 117 |
|---|---|---|
| Pricing Unit: | lb. | |
| Standard/Grade: | Select No. 1 | |
| | Moderately thick slice | |
| | Oscar Mayer item 2040 or equal | |
| Weight Range: | 14–16 slices per lb. | |
| Packaging: | 2/10 lb. Cryovac packed | |
| Container Size: | Not to exceed 20 lb. | |
| Intended Use: | Bacon, lettuce, and tomato sandwiches | |
| Other Information: | Flat packed on oven-proofed paper | |
| | Never frozen | |
| **Product Yield**: | 60% Yield (usable amount) | |

**Figure 3.2** Sample Product Specification: Bacon (slice)

A properly prepared product spec includes the following information:

1) Product name or specification number
2) Pricing unit
3) Standard or grade
4) Weight range/size
5) Processing and/or packaging
6) Intended use

### Product Name (Specification Number)

Product names are not always self-explanatory. For example, chickpeas and gar-banzo beans are the same item, but they may be packaged under either name. The same is true for sweet potatoes and yams (although they are technically different items). Similarly, large shrimp and prawns may mean the same product to some vendors but not to others. Therefore, a product's name on a specification must be detailed enough to clearly and precisely identify the item an operator wants to purchase.

When developing product specifications, many operators find it helpful to assign a number and a name to the item. This can be useful when, for example, several forms of the same ingredient or menu item are purchased.

To illustrate, consider a deli-restaurant that may use 5–10 different types of bread depending on its intended use. Each type of bread may be given a specific name and spec number. The same may be true with other needed items such as cheese that is available in several forms (e.g., block, sliced, or shredded) and types (e.g., Colby, Cheddar, or Swiss). Note that, in Figure 3.2, the desired type of bacon is given a name and a specification number (Spec #117).

### Pricing Unit

A product's **pricing unit** may be pounds, quarts, gallons, cases, or other commonly used measurement unit in which the item is sold. Many pricing units such as pounds and quarts are easily understood.

However, some pricing units are less well known. Parsley, for example, is typically sold in the United States by the bunch, and grapes are

**Key Term**

**Pricing unit:** The measure used to establish the selling price of a product. For example, pound, quart, case, bushel, or bag.

sold by the lug. Other commonly used pricing units include cartons, trays, bushels, and flats. To avoid confusion, operators should insist that their vendors supply them with the definitions of each pricing unit upon which they base their selling prices.

**Standard or Grade**

Many food items are sold with varying degrees of quality. The U.S. Department of Agriculture (USDA), the Bureau of Fisheries, and the Food and Drug Administration have developed standards (grades) for many food items. In addition, grading programs are in place for many commonly used foodservice items. Trade groups such as the National Association of Meat Purveyors publish item descriptions for many of these products. Consumers also are aware of many of these distinctions. One good example of this is the USDA's grades for beef used in most foodservice operations. In descending order of quality, these are:

✓ Prime
✓ Choice
✓ Select
✓ Standard (Commercial)

In some cases, operators may prefer a spec identifying a particular product's brand name (e.g., Minor's beef base or A-1 Steak Sauce) or a specific point of origin (e.g., Maine lobster or Idaho potatoes) as an alternative or supplement to a product standard or grade.

---

**Find Out More**

Food service operators seeking to buy high-quality products have an important ally in the USDA, which has the quality grade marks usually seen on beef, lamb, chicken, turkey, butter, and eggs. For many other products such as fresh and processed fruits and vegetables, the grade mark is not always visible on the retail product.

With many commodities, the grading service is used by wholesalers, and the final packaging may not include the grade mark. However, quality grades are widely used—even if they are not prominently displayed—as a "language" among traders. They make business transactions easier whether they are local or made over long distances. Foodservice operators, as well as those involved in the marketing of agricultural products, benefit from the greater efficiency permitted by the availability and application of grade standards.

The USDA's Agricultural Marketing Service (AMS) currently grades fresh fruits and vegetables, processed fruits and vegetables, poultry, eggs, livestock and meat, and dairy products. To learn about the specific characteristics of food items that result in their various quality grades, enter "grades" in the search bar of USDA.gov and review the results.

---

### Weight Range/Size

Weight range or size is important when purchasing meats and some fish, poultry, fruits, and vegetables. In the standardized recipe example of roast chicken (see Chapter 1, Figure 1.5), the chicken quarters should be from chickens weighing 3.0–3.50 pounds. Note: The weight range will make a big difference in portion cost and cooking times compared to, for example, chickens weighing more or less than this weight.

When products require a specific trim or maximum fat covering, this should also be designated. For example, a 10-ounce strip steak may be ordered with a maximum tail of 1 inch and fat covering of ½ inch. Other examples of items that may require an exact size rather than a weight range include four-ounce hamburger patties, 16-ounce T-bone steaks, and ¼-pound hot dogs.

**Count** is a purchasing term used to designate product size. For example, 16- to 20-count shrimp means that, for this size shrimp, 16 to 20 individual shrimp are contained in one pound. Similarly, 30- to 40-count shrimp means there are 30 to 40 of this size shrimp in one pound. In addition to seafoods such as shrimp

**Key Term**

**Count (product):** A number used to designate product size or quantity when purchasing food items.

and scallops, many fruits and vegetables are sold by count. In general, the larger the count, the smaller the size of the individual food items.

### Processing and/or Packaging

Processing and packaging refer to a product's state when purchased. Peaches, for example, may be purchased fresh, dried, canned, or frozen. Each form specifies an appropriate price when it is purchased.

Food can come packed in numerous forms and styles including slab packed, layered cell packed, fiberboard divided, shrink-wrap packed, individually wrapped, and bulk packed. The scope of this book does not detail all varieties of food processing and packing styles available, but it is important for operators to have some knowledge about them. Vendors chosen by the operator should provide information about all processing and packaging alternatives offered for the products they sell.

---

**Technology at Work**

"Farm to Fork" (F2F) (or "Farm to Table") is a term of growing importance to foodservice operators and their guests. F2F refers to the path food follows from those who grow or raise it to those who will prepare and serve it. Ideally, this path is short to maximize freshness, to minimize health risks, and to be environmentally friendly.

In the future, advancements in technology will likely have greater and greater impact on the F2F process. Sensors on fields and crops can provide

---

farmers literally granular data points on soil conditions; provide detailed information about wind, fertilizer requirements, water availability, and pest infestations; as well as indicate peak harvesting periods.

Data analytics can help prevent spoilage by moving harvested products faster and more efficiently. In addition, radio frequency identification (RFID)-based traceability systems can provide a constant data stream on farm products as they move in the supply chain from harvesting to composting after use. Improvements in transportation tracking and lower-cost packaging costs can also translate into lower prices charged to foodservice operators.

Locally grown and organic products also help support the same local economies from which many foodservice operators draw their customers. As a result, buying locally is an option that should be thoroughly examined for benefits to their communities, the environment, customers' health, and profit levels.

To learn more about this popular purchasing approach, enter "Farm to Fork trends," or "Farm to Table trends," in your favorite browser and view the results.

### Intended Use

In foodservice operations, different types of the same item are often used in a variety of ways. Consider the operator who uses strawberries. Perfectly colored and shaped large berries are best for chocolate-dipped strawberries served on a buffet table. Less-than-perfect berries, however, may cost less and be perfectly acceptable for sliced strawberries used to top strawberry shortcake. Frozen berries may make a good choice for a baked strawberry pie and may even be of lower cost.

Breads, dairy items, apples, and other fruits are additional examples of foods that are available in several forms. The best form of a food product is not necessarily the most expensive form. Rather, it is the form that is best for a product's intended use.

When clear-cut specifications describing what they should buy are in use, cost-conscious operators must then turn their attention to how much they should buy.

## Determining Needed Product Quantities

As they plan their purchases, foodservice operators must estimate their future product needs. Operators with accurate sales histories must use them to forecast future sales, and guest counts (see Chapter 2) will have records of their guest's past purchases. These records will be held in an operation's point-of-sales (POS) system and an analysis of these historical records allows operators to know how much of each product to be purchased will be needed for service to future guests.

As they work to determine the quantity of products needed, operators undertake two important tasks. These are forecasting menu item sales and establishing desired inventory levels.

## Forecasting Menu Item Sales

To illustrate the process of forecasting menu item sales, assume that, based on prior sales histories, an operator forecasts that 300 guests will visit their operation for lunch next Monday. Assume further that the operation sells only three entrée items: roast chicken, roast pork, and roast beef. A question the operator must consider is, "How much chicken, pork, and beef must be purchased so we do not run out of any item nor have too many of each item remaining after the lunch period?"

If the operation ran out of one of the three menu items, guests who wanted that item would likely be upset. They might even leave the operation (perhaps taking dining companions with them who would have ordered other items produced by the operation). Producing too much of any one item results in unsold items and causes product costs to rise unnecessarily unless these remaining items could be sold at full price later.

In this situation, it would be unwise to buy the products needed to produce 300 portions of each item. If it did, the operation would not run out of any one item (each of the 300 estimated guests could order the same item, and the operation would still have produced a sufficient quantity). However, it would also have 600 portions (900 portions produced − 300 portions sold = 600 portions remaining) left over when the lunch period ended.

What this operator would like to do is to buy the products that allow their production staff to prepare the "right" amount of each menu item. Note: The "right" number of servings minimizes the chances of running out of an item before lunch is over. The "right" number also reduces the chance of having an excessive amount unsold when the meal period ends.

The answer to the question of how many servings of roast chicken, pork, and beef should be purchased and prepared lies in accurate menu item sales forecasting.

Returning to the three-item menu example, assume the operator used POS system data to record last week's sales of menu items on a form like the one presented in Figure 3.3.

A review of the Figure 3.3 data indicates an estimated 300 guests for next Monday makes good sense because the weekly sales total last week (1,500 guests) averages 300 guests per day (1,500 guests served/5 days = 300 guests served per day).

In this example, last week the operation sold an average of 73 roast chicken (365 sold/5 days = 73 per day), 115 roast pork (573 sold/5 days = 115 per day), and

Date: 1/2–1/6 Menu Items Sold

| Menu Item | Mon | Tues | Wed | Thurs | Fri | Week's Total | Average Daily Sales |
|---|---|---|---|---|---|---|---|
| Roast Chicken | 70 | 72 | 61 | 85 | 77 | 365 | 73 |
| Roast Pork | 110 | 108 | 144 | 109 | 102 | 573 | 115 |
| Roast Beef | 100 | 140 | 95 | 121 | 106 | 562 | 112 |
| Total | 280 | 320 | 300 | 315 | 285 | 1,500 | 300 |

**Figure 3.3** Menu Item Sales History

112 roast beef (562 sold/5 days = 112 per day) portions per day.

An operator must know the average number of people selecting a given menu item, and they must also know the total number of guests who made the selections. Then they can compute each menu item's **popularity index**: the proportion of total guests choosing a given menu item from a list of alternative menu items.

If it is assumed that future guests will select menu items in a manner like that of past guests, the information can improve sales forecasts using the following formula:

$$\frac{\text{Total Number of a Specific Menu Item Sold}}{\text{Total Number of All Menu Items Sold}} = \text{Popularity Index}$$

**Key Term**

**Popularity index:** The proportion of guests choosing a specific menu item from a list of alternative menu items. The formula for a popularity index is:

$$\frac{\text{Total Number of a Specific Menu Item Sold}}{\text{Total Number of All Menu Items Sold}}$$
$$= \text{Popularity Index}$$

The popularity index for roast chicken sold last week would be 24.3% (365 roast chickens sold/1,500 total guests = 0.243, or 24.3%). Similarly, 38.2% (573 roast pork sold/1,500 total guests = 38.2%) of guests preferred roast pork, and 37.5% (562 roast beef sold/1,500 total guests = 37.5%) selected roast beef.

When an operator knows the number of menu items future guests will select, they are better prepared to make good decisions about the products they must purchase and the quantity of each menu item to produce. The basic formula for individual menu item forecasting based on the item's sales history is:

Number of Guests Expected × Item Popularity Index
    = Predicted Number to Be Sold

In this example, Figure 3.4 illustrates the operator's best estimate of what 300 guests are likely to order:

| Menu Item | Guest Forecast | Popularity Index | Predicted Number to Be Sold |
|---|---|---|---|
| Roast Chicken | 300 | 0.243 | 73 |
| Roast Pork | 300 | 0.382 | 115 |
| Roast Beef | 300 | 0.375 | 112 |
| Total | | | 300 |

**Figure 3.4** Forecasted Item Sales

The predicted number to be sold is the quantity of a specific food or beverage item likely served given a reasonable estimate of the total number of guests expected to be served.

The best operators know which menu items their guests are likely to select, and they use standardized recipes (see Chapter 1) to clearly indicate the ingredients needed to produce the items. The reason: They can prepare an accurate "shopping list" of products to be purchased or be available in inventory.

### Establishing Inventory Levels

It is theoretically possible to forecast the future sales of each menu item offered for sale and to utilize standard recipes to identify the sum of the specific ingredients needed to make the required items. However, this is not the most efficient way for operators to determine the quantity of products to be ordered. Rather, operators should establish inventory levels that allow them to have the required amount of needed products on hand and available for production staff.

A goal of proper inventory management is to provide appropriate **working stock**: the amount of an ingredient that will be used before purchasing that item again. In addition, a reasonable **safety stock** should be established: an extra amount of that ingredient an operator decides to keep on hand to meet higher-than-anticipated demand.

It is important to recognize that demand for a given menu item can fluctuate greatly between supplier deliveries, even when the deliveries are frequent. With too little inventory on hand, an operator may run out of products and, therefore, potentially reduce guest satisfaction and

**Key Term**

**Working stock:** The quantity of goods from inventory reasonably expected to be used between deliveries.

**Key Term**

**Safety stock:** Additions to working stock, held as a hedge against the possibility of extra demand for a given menu item. Note: This helps reduce the risk of being out of stock on any given inventory item.

increase employee frustration. If too much inventory is kept on hand, however, waste, theft, and spoilage can become excessive. The ability to effectively manage the inventory process is one of the best skills a foodservice operator can acquire.

The actual quantity of products operators should buy and keep in inventory will vary based on a number of key factors including:

✓ Storage capacity
✓ Item perishability
✓ Vendor delivery schedule
✓ Potential savings from increased purchase size
✓ Operating calendar
✓ Relative importance of stock outages

### Storage Capacity

Inventory items must be purchased in quantities that can be adequately stored and secured (see Chapter 5). However, kitchens may often lack adequate storage facilities. This may mean that more frequent deliveries are needed to include less of each product on hand than would otherwise be desired. When storage space is too great, however, the tendency of some operators is to fill the storage space. It is important that this is not done because increased inventory of items can lead to greater spoilage and loss due to theft. Moreover, large quantities of goods in storage may send a message to production staff that there is "plenty" of everything. This may then result in careless use of valuable and expensive products. It is also unwise to overload refrigerators or freezers because these units will then operate less efficiently, and stored items may be difficult to locate and might suffer loss of quality.

### Item Perishability

If all food products retained the same level of freshness, flavor, and quality while in storage, operators would have little difficulty in maintaining proper product inventory levels. The **shelf life** of food products, however, varies greatly.

For example, the shelf life of many fresh produce, seafood, and meat items may be several days, while the shelf life of some frozen and canned items may be several weeks or months.

**Key Term**

**Shelf life:** The amount of time a product held in inventory retains peak freshness and quality.

Since different food items have varying shelf lives, foodservice operators must balance the need for a particular product with the optimal shelf life of that product. Serving items that are "too old" is a sure way to produce guest complaints. In fact, one of the quickest ways to determine the overall effectiveness of a foodservice operator is to "walk the boxes." Note: This term means to take a tour of an operation's storage areas, and to then address problems that are noted.

If many products, especially those in refrigerated areas, are moldy, soft, over-ripe, or rotten, this indicates that designated employees do not understand proper inventory levels based on the shelf lives of stored items. It is also a sign that menu item sales forecasting methods are either not in place or are not working well.

---

### Find Out More

The best foodservice operators want to serve all the ingredients they utilize at the ingredients' peak quality level. Doing so means carefully monitoring their purchasing and storage procedures as well as inventory levels to assure that food and beverage products are used within their appropriate shelf life.

The shelf life of different foods can vary tremendously. Sometimes, a visual inspection of a product can determine whether it is at or beyond its peak quality level. In other cases, determining the appropriate quality level can be more complex. The United States Department of Agriculture (USDA) provides food-service operators with guidance on peak quality levels for various types of foods. Consider, for example, this important canned food-related information posted on the USDA website:

*Canned Foods: Dates on cans indicate peak quality as determined by the manufacturer. So don't automatically pitch a can with an expired date. You can safely keep commercially canned foods longer than their dates. Low-acid foods (such as canned meat, poultry, fish, stew, soups, green vegetables, beans, carrots, corn, peas, potatoes, etc.) can be stored for two to five years; high-acid foods (e.g. canned juices, fruit, pickles, sauerkraut, tomatoes, tomato soup), for 12-18 months.*[1]

Knowing appropriate shelf lives of various foods can help operators make better purchasing decisions. To learn more about USDA recommendations or the shelf life of items to be held in inventory, enter "USDA recommended shelf life for common foods" in your favorite search engine and review the results.

---

### Vendor Delivery Schedule

It is the fortunate foodservice operator whose business is in a large city with many suppliers, some of whom may offer the same products and all of whom would like to have the operator's business. In many cases, however, an operator will not have the luxury of daily delivery. An operation may be too small to warrant such

---

1 https://www.usda.gov/media/blog/2014/08/19/save-money-knowing-when-food-safe#:~:text=Low%2Dacid%20foods%20(such%20as,always%20to%20observe%3A%20infant%20formula. Retrieved May, 1, 2023.

frequent stops by a supplier and/or the operation may be in a remote location where daily delivery is not offered by potential vendors.

Consider, for example, the difficulty confronting an operator of a foodservice facility on an offshore oil rig operating in the Gulf of Mexico. Clearly, a supplier willing to provide daily doughnut delivery is going to be hard to find! Also, it is important to remember that the costs to the supplier of frequent deliveries will be reflected in the prices the supplier will charge.

---

**What Would You Do? 3.1**

"I don't' believe this," said Janie Collins to Mike Ventrella. "We talked about this just last month, and now it has happened again. What is it going to take to get you and Kaminsky Brothers on the same page?"

Janie, the kitchen manager at Albarran's Mexican Grill, was talking to Mike, the Grill's manager and the individual in charge of purchasing.

"You told me you needed 48 Hass avocados for Friday night," replied Mike. "It's Friday night, and there are 48 Hass avocados in the walk-in, just like you wanted. They were delivered this morning!"

"No!" replied Janie, "There are 48 avocados in the walk-in that are as hard as rocks. They won't be ripe and ready to use to make our guacamole for three or four days at least! And I need them tonight!"

**Assume you were Mike. Who do you think is responsible for ensuring that "ripe" avocados are ready for kitchen production use at the right time? How could you work with the Kaminsky Brothers produce company to ensure that this kind of problem does not happen in the future?**

---

### Potential Savings from Increased Purchase Size

Sometimes operators may find they can realize substantial savings by purchasing needed items in large quantities. Doing so makes good sense if the total savings outweigh the added costs of receiving and storing the larger quantity. It is important to recognize, however, that there are costs associated with extraordinarily large purchases. These may include storage costs, spoilage, deterioration, insect and rodent infestation, and/or theft.

Generally, operators should determine their ideal product inventory levels and then maintain their stock within those ranges. Only when the advantages of placing an extraordinarily large order are very clear should such a purchase be undertaken.

### Operating Calendar

When an operation serves meals seven days a week to a relatively stable number of guests, the operating calendar makes little difference to inventory-level decision making. If, however, the operation opens on Monday and closes on Friday for two days, as is the case in many school foodservice accounts, the operating calendar plays a large part in determining desired inventory levels. In general, an operator who is closing either for a weekend or for a season (as in the operation of a summer camp or an operation located in a seasonal resort area) should attempt to greatly reduce overall **perishable inventory** levels as the closing period approaches.

**Key Term**

**Perishable inventory:** Inventory items that possess relatively short shelf lives.

### Relative Importance of Stock Outages

In many foodservice operations, not having enough of a single food ingredient or menu item is unimportant . In other operations, the shortage of even one menu item might spell disaster. For example, it may be all right for the local French restaurant to run out of one of the specials on Saturday night. However, it is difficult to imagine the problem of the McDonald's restaurant operator who runs out of French-fried potatoes on that same Saturday night!

For small operators, a mistake in the inventory level of a minor ingredient that results in an outage might be corrected by a quick run to the local grocery store. For larger facilities, such an outage may well represent a substantial loss of sales or guests' goodwill.

In the foodservice industry, when an item is no longer available on the menu, operators "86" the item, a reference to restaurant slang originating in the early 1920s (86 rhymed with "nix," a Cockney term meaning "to eliminate"). If an operator finds they must "86" too many items on any given day, the reputation of the operation will likely suffer.

A strong awareness and knowledge of how critical outage factors helps operators to determine their appropriate product inventory levels. A final word of caution is, however, necessary. The foodservice operator who is determined to never run out of anything must be careful not to set inventory levels so high that costs to the operation are more than if realistic levels were maintained. Note: Specific details on how to establish and maintain appropriate inventory levels for products in storage is addressed in detail in Chapter 5.

## Placing the Purchase Order

After an operator determines the amount of product to be purchased, it is still necessary to address several key areas:

✓ Choosing suppliers
✓ Understanding suppliers
✓ Recording purchases

## Choosing Suppliers

The best foodservice operators want to establish a partnership arrangement with their chosen suppliers. Some observers believe a true "partnership" with an operator implies sharing in the operation's financial risk and reward. However, a partnership can also involve businesses with joint interests who work together as they maximize efforts to mutually benefit from their interactions. This use of the term "partnership" relates to foodservice suppliers because, when the supplier's customer (a foodservice operation) succeeds, the supplier succeeds as well. For foodservice operators, selecting their supplier partners is a very important task.

Many desired characteristics of high-quality suppliers are easy to define, but others may also be unique to the needs of a specific foodservice operation. In nearly all cases, however, professional operators recognize that good suppliers are those who:

✓ Consistently provide the quality of products specified by the operator
✓ Offer products at a reasonable price
✓ Meet product delivery schedules
✓ Provide useful support services
✓ Take ownership of problems when they occur and respond to buyers' needs
✓ Inform the buyer about any order/delivery problems in a timely manner
✓ Enjoy a stable financial position that is necessary for the supplier to remain in business
✓ Are mutually interested in providing value to customers and do so, in part, by suggesting how costs can be reduced without sacrificing quality standards
✓ Have similar values about ethical business relationships
✓ Emphasize product and service quality
✓ Employ a highly motivated workforce to minimize problems from high employee turnover rates and/or union-related work stoppages
✓ Have a genuine interest in helping foodservice operators achieve their goals
✓ Are readily accessible so communication between an operator and the supplier is "easy"

When choosing their suppliers, some operators may conduct on-site inspections of a potential supplier's facilities to observe factors such as work methods, cleanliness and other food safety practices, and the general organization of the facility. The condition of transport equipment is also important when, for example, highly perishable fresh produce, meats and seafood, and dairy products must be delivered during warm weather.

Some operators may actually "score" potential suppliers on a variety of factors and then select those suppliers who achieve the highest scores. Figure 3.5 is an example of a pre-selection scoring system that foodservice operators may use to determine if a supplier is one whom operators might consider when selecting suppliers.

| Supplier Name: | Contact Information: |
| --- | --- |
| _____ | |
| Products Provided: _____ | Representative: _____ |
| | Telephone: _____ |
| | E-mail: _____ |
| | Address: _____ |
| | _____ |
| | _____ |

**Sources of Information about Supplier (check all that apply):**

- ☐ Interviews with supplier references
- ☐ Distributor's sales representative
- ☐ Sales Manager
- ☐ Other interviews:_____

- ☐ On-site visit
- ☐ Trade publications
- ☐ Internet marketing information
- ☐ Other: _____

| Evaluation Factors | Unacceptable | Acceptable | Comments |
| --- | :---: | :---: | --- |
| Quality (Follows Standards) | ☐ | ☐ | _____ |
| Service Procedures | ☐ | ☐ | _____ |
| Service Philosophy | ☐ | ☐ | _____ |
| Management Systems | ☐ | ☐ | _____ |
| Facilities/Delivery Equipment | ☐ | ☐ | _____ |
| E-Commerce Applications | ☐ | ☐ | _____ |
| Financial Stability | ☐ | ☐ | _____ |
| Reputation | ☐ | ☐ | _____ |
| Information (Technical Support) | ☐ | ☐ | _____ |
| Input from others | ☐ | ☐ | _____ |
| Sanitation/Food Safety (if applicable) | ☐ | ☐ | _____ |
| Current Customer Recommendations | ☐ | ☐ | _____ |
| Experience (years in business)_____ | ☐ | ☐ | _____ |
| **Total Points** | 0 | _____ | |

Other Information:_____
_____
_____

Supplier Selection Recommendation/Decision
_____
_____
_____

Buyer/Operator:_____  Date: _____

**Figure 3.5**  Supplier Pre-Selection Form

In some cases, foodservice operators (and especially those representing smaller businesses) may elect to become their own suppliers by frequenting wholesale buying clubs such as Sam's Club or Costco, or by purchasing one or more products at "open to the public" restaurant supply businesses.

## Understanding Suppliers

Many operators must decide whether to buy specific items from one or many suppliers. In general, the more suppliers used, the more time operators must spend in ordering, receiving, and paying vendor invoices. Some operators, however, fear that if they give all their business to only one supplier, their costs may rise because of a lack of competition.

However, the likelihood of this occurring is extremely small. Consider, for example, that foodservice operators are unlikely to take advantage of their best guests. In fact, they might offer additional services not available to occasional guests. Likewise, many suppliers also make extra efforts to help operators who do most of their buying from that supplier. In fact, it makes good business sense for the vendor to do so.

Unfortunately, too little has been shared in the field of foodservice operations about managing costs through cooperation with suppliers. An operation's suppliers can be some of an operator's most important allies in controlling costs. Operators who determine their suppliers only because of the low prices offered often discover they receive only the purchased products, while their competitors may be buying more than just food! Or, as one food salesperson said when asked why their company should be selected as the primary food vendor for a business, "With my products, you also get my knowledge!"

Just as the best foodservice operators know their guests appreciate high-quality personal service levels, the best vendors also know that operators appreciate high-quality customer service. Suppliers to many foodservice operations can be of immense value in ensuring a steady supply of quality products at a fair price if foodservice operators understand several facts about suppliers.

1) Suppliers Have Many, Not Just One, Prices

   Unlike the foodservice business that generally charges the same price to every buyer, suppliers have a variety of prices based on the customer being quoted product prices. The price an operator will pay is often based upon a variety of factors including the total amount of food and other products purchased and the promptness with which operators pay their bills.

2) Suppliers Reward Volume Buyers

   It is in the best interests of a supplier to give a better price to a high-volume buyer. The cost to deliver a $2,000 food order to a restaurant is not that much different from the cost to deliver a $200 order. Each of these orders require one truck and one driver, and operators who decide to concentrate their business in the hands of fewer suppliers will generally pay lower prices.

3) Cherry Pickers Are Typically Serviced Last

   **Cherry picker** is the term used by vendors and other suppliers to describe the customers who get bids from multiple vendors and then buy only the items that each vendor has "on sale" or for the lowest price.

**Key Term**

**Cherry picker:** A foodservice operator who buys only those items from a supplier that are the lowest in price among the supplier's competition.

If an operator buys only a vendor's "on sale" items, the supplier will usually respond by providing limited service. It is a natural reaction to the foodservice operator's failure to consider varying service levels, long-term relationships, dependability, or any other supplier characteristic except least expensive price.
4) Slow Pay Means High Pay

In most cases, operators who are slow to pay their bills will find that vendors charge them more for products than they charge other operators who pay their bills in a timelier manner.

Carefully selecting and understanding suppliers is a key step in the purchasing process. Regardless of their final supplier selection decisions, as operators place their orders and/or make their self-purchases, they must also carefully make and retain an accurate record of their costs. The reason: Records will be immensely valuable for several cost control-related reasons.

## Recording Purchases

Regardless of the food or beverage items being purchased, all operators should use a purchase order (PO) to formally record the items ordered from suppliers. A formal PO may include a variety of product information, but it must always include the quantity ordered and the price quoted by the supplier. When ordering products, operators may choose several ways to tell suppliers what they want to buy. These methods may include ordering via the Internet, by telephone, or by placing orders in person.

Regardless of the communication method utilized, it is critical that there is a written or printable PO because it serves as the formal record of what has been purchased.

An operator's POs can be simple or complex, but they should include the following information:

1) Vendor contact information
2) Purchase order number
3) Date ordered
4) Delivery date requested
5) Name of person who placed order
6) Names of ordered items
7) Item specification #, if appropriate
8) Quantity ordered
9) Quoted price (per unit)
10) Total PO cost
11) Delivery instructions (if applicable)
12) Comments section (optional)

Figure 3.6 shows a sample PO for various produce items ordered from Skippy's Produce by Jennifer James, the operator of the Highlander Restaurant.

Regardless of the format used, a PO must always include the specific items ordered, the quantities of ordered items, and the ordered items' quoted prices.

Highlander Restaurant Purchase Order

**Vendor:** Skippy's Produce     **Purchase Order #:** 256

**Vendor's Address:** 123 Anywhere Street, Any City, Any State     **Delivery Date:** 1/18

**Vendor's Telephone #:** 1-800-999-0000     **Vendor's E-mail:** skippyproduce@isp.org

| Item Purchased | Spec # | Quantity Ordered | Quoted Price | Extended Price |
|---|---|---|---|---|
| 1. Bananas | 81 | 30 lb. | $0.44 lb. | $ 13.20 |
| 2. Parsley | 107 | 4 bunches | $0.80/bunch | $ 3.20 |
| 3. Oranges | 101 | 3 cases | $31.50/case | $ 94.50 |
| 4. Lemons | 35 | 6 cases | $29.20/case | $175.20 |
| 5. Cabbage | 85 | 2 bags | $13.80/bag | $ 27.60 |
| 6. | | | | |
| 7. | | | | |
| 8. | | | | |
| 9. | | | | |
| 10. | | | | |
| Total | | | | $313.70 |

**Order Date:** 1/15
**Order Confirmed Date:** 1/16     **Comments:** Please notify in advance if there are product outages.

**Ordered By:** Jennifer James

**Received By:** _____     **Order Transmitted by:** Jennifer James

**Delivery Instructions:** After 1:00 p.m. only

**Figure 3.6** Sample Purchase Order

---

**Technology at Work**

A professional purchasing program is essential in the foodservice business. All foodservice operators must ensure that they have the right ingredients, in the right amounts, for the right prices, and at the right times.

When a professional purchasing program is in place, operators can purchase high-quality ingredients from the best available suppliers and avoid excessive inventory levels. Excess inventory is a problem in all businesses but especially so in foodservice because excess products held in storage can spoil or otherwise deteriorate over time.

In many cases, suppliers' own online purchasing programs provide operators with an electronic record of their purchases. However, few operations buy all products from the same supplier, so most operators must also have their own purchasing software and recordkeeping ability.

Fortunately, several good companies produce purchasing software designed specifically for the restaurant industry. These programs help operators manage the purchasing task and help them determine how much of each ingredient they need and when they will need it.

To learn more about these useful programs, enter "restaurant purchasing software" in your favorite search engine and review the results.

After a PO is submitted to a supplier, the ordered items must be properly received when delivered. A specially designated receiving clerk (in a large operation) may perform the receiving function, but in smaller operations an owner, a manager, a supervisor, or a designated staff member likely is responsible. The proper receiving of purchased items is so important to a foodservice operator's cost control efforts that it will be the main topic of the next chapter.

### What Would You Do? 3.2

"So you would save almost 60 cents a pound," said Ricky Watts, "with as many pounds of ground beef as you buy every week, those savings will really add up!"

Ricky Watts, a sales representative with White Swan Foods, was talking to Mandy Davis, owner of Mandy's Burger Palace.

The Burger Palace was a 150-seat operation that was known locally for serving the best burgers in the area. Business was good, and the customer reviews posted online were consistently excellent.

Mandy's burgers were made with fresh (never frozen!) all-beef patties, and she advertised that fact widely.

"Yes," replied Mandy, "I know the cost per pound is less, but it's a frozen patty."

"Yes, they are frozen," replied Ricky, "but all the big chains use frozen burger patties now and it's no problem. And in addition to being less expensive per pound, you don't run the risk of any spoilage. It's a win-win!"

**Assume you were Mandy. What cost-related issues are in play as you consider this purchasing decision? What guest service and marketing-related issues would be applicable as you consider this purchasing decision?**

## Key Terms

| | | |
|---|---|---|
| Purchasing | Product yield | Safety stock |
| Quality | Pricing unit | Shelf life |
| Purchase order (PO) | Count (product) | Perishable inventory |
| Product specification (spec) | Popularity index | Cherry picker |
| | Working stock | |

---

**Operator's 10-Point Tactics for Success Checklist**

---

Evaluate your need for, and the current status of, each of the following operational tactics. For those tactics you think are important, but not yet in place, develop an action plan for its implementation including who will be responsible for the tactic's completion and the target date by which it should be completed.

| | | | | If Not Done | |
|---|---|---|---|---|---|
| Tactic | Don't Agree (Not Done) | Agree (Done) | Agree (Not Done) | Who Is Responsible? | Target Completion Date |
| 1) Operator recognizes the objectives of a professional purchasing program. | ___ | ___ | ___ | | |
| 2) Operator can identify the steps required to implement a professional purchasing program. | ___ | ___ | ___ | | |
| 3) Operator can name the information contained in a properly prepared product specification (spec). | ___ | ___ | ___ | | |
| 4) Operator recognizes the necessity of forecasting menu item sales when determining product need levels. | ___ | ___ | ___ | | |
| 5) Using sales histories, operator can calculate the popularity index of all menu items offered for sale. | ___ | ___ | ___ | | |

|  | | If Not Done | | |
| Tactic | Don't Agree (Not Done) | Agree (Done) | Agree (Not Done) | Who Is Responsible? | Target Completion Date |
|---|---|---|---|---|---|
| 6) Operator recognizes the necessity of establishing desired inventory levels when determining how much product to buy and knows the key factors that impact inventory levels. | ―― | ―― | ―― | | |
| 7) Operator can state the characteristics of high-quality suppliers in the foodservice industry. | ―― | ―― | ―― | | |
| 8) Operator understands the purpose and use of a supplier pre-selection form. | ―― | ―― | ―― | | |
| 9) Operator recognizes the primary purpose of placing orders using a formal purchase order (PO). | ―― | ―― | ―― | | |
| 10) Operator can state the specific information that must be contained on a formal purchase order (PO). | ―― | ―― | ―― | | |

# 4

# Cost Control in Receiving

**What You Will Learn**

1) The Importance of Professional Receiving
2) The Essentials of Professional Receiving
3) How to Evaluate Key Characteristics of Delivered Items
4) How to Properly Maintain Receiving Records

**Operator's Brief**

In this chapter, you will learn that properly receiving ordered products is a multi-step process, and each step must be carefully managed. Professional receiving procedures in a foodservice operation are critical to controlling receiving-related costs. There are basics that you must implement as you develop a professional receiving system. These essentials are a proper location, proper tools and equipment, and proper delivery schedules.

To ensure that only appropriate products are accepted for delivery, your receiving clerks must be well-trained. Clerks must check incoming products for their proper weights to verify that the operation is charged only for the product weight delivered—not for additional costs including mixing high- and low-cost items in one package.

The proper counting of products is as important as their proper weighing. A reason: Suppliers typically make more invoicing mistakes in *not* leaving products than they do in making excessive deliveries. Shorted products must be made known to you and delivery personnel so invoices can be properly adjusted (using credit memos) and so shorted (missing) products can be re-ordered as needed.

Incoming products must be inspected to ensure they comply with your established product specifications (specs) and that the prices charged for them are correct. If incoming products are not compared to product specs, substandard products may be accepted. Similarly, if incoming product prices are not compared to those initially agreed upon, your operation may be over-charged (or undercharged), and this must be avoided.

Finally, in this chapter, you will learn that proper recordkeeping is essential to professional receiving. Records that must be kept include invoices provided by suppliers upon product delivery and receiving-related records produced by your operation.

---

**CHAPTER OUTLINE**

The Importance of Controlling Receiving Costs
Receiving Essentials
    Proper Location
    Proper Tools and Equipment
    Proper Delivery Schedules
Key Areas of Product Verification
    Weight
    Quantity
    Quality
    Price
Receiving Records
    Supplier-Produced Delivery Records
    Operation-Produced Delivery Records

---

## The Importance of Controlling Receiving Costs

After a purchase order (PO) has been submitted to a supplier, a foodservice operator must ensure the operation is ready to accept (receive) the order. In a large foodservice operation, receiving tasks may be performed by a designated receiving clerk. However, in most smaller operations, the responsibilities of receiving fall upon the operation's owner, manager, or other designated individual.

In all cases, however, it is a good idea for an operation's manager to establish the purchasing and receiving functions so one individual places orders, and a different individual is responsible for verifying delivery and acceptance of the orders. If this is not done, the potential for purchasing fraud or theft can be substantial.

In fact, foodservice **auditors** who discover purchasing- and receiving-related fraud frequently find that a buyer ordered a product, signed for its acceptance, and authorized payment for it when, in fact, no product was ever delivered! In this case, the buyer could be receiving cash payments from the supplier without the business owner's

**Key Term**

**Auditor:** An individual or entity responsible for reviewing and evaluating proper operating procedures in a business.

knowledge. In other cases, a buyer may purchase items, accept delivery of the items, but then remove the items from inventory for their own personal use.

When purchasing duties are split among two or more individuals in the purchasing chain, these types of fraud are much less likely to happen. If it is not possible to have more than one person involved in the buying process, the work of buyers and receiving clerks must be carefully monitored by business owners to prevent fraud.

Even when no fraud is occurring proper receiving methods must be used to ensure that only delivered products are paid for, and that all delivered products have met pre-established specifications (specs) (see Chapter 3). Finally, proper receiving records must be in place to ensure that the operation will pay only the prices that were agreed upon when purchase orders (POs) were initially submitted. If significant errors occur in any of these procedures, the negative effects on product costs can be substantial.

Figure 4.1 shows the steps to be taken to ensure proper receiving in a foodservice operation, and each of these will be addressed in detail in this chapter.

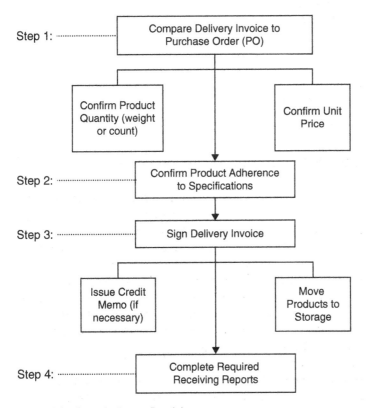

**Figure 4.1** Steps in Proper Receiving

# Receiving Essentials

Professional receiving procedures in a foodservice operation are essential to controlling receiving-related costs. The development of a proper receiving program requires that some essentials be in place. For all foodservice operations, these essentials include:

✓ Proper Location
✓ Proper Tools and Equipment
✓ Proper Delivery Schedules

## Proper Location

An operation's "back door" that is most often used for receiving is frequently no more than that: an entrance to the kitchen. In fact, the receiving area must be adequate to handle the job of receiving or product loss and inconsistency can result.

First, the receiving area must be large enough to allow for properly checking products delivered against both the **delivery invoice** and the PO (the buyer's record of what was originally ordered).

**Key Term**

**Delivery invoice:** A supplier's record of what is being delivered and at what cost.

In addition to the space required to count and weigh incoming items, accessibility to equipment required to move products to their proper storage area and to dispose of excess packaging is important. A location near refrigerated storage areas is desirable for maintaining deliveries of refrigerated and frozen products at their optimal temperatures.

The receiving area must stay free of trash and clutter. The reason: Some items might be easy to hide, and delivered food items may be taken home at the end of a dishonest employee's shift. It is also important to remember that the delivery person can be a potential thief. Most suppliers are careful to screen their delivery personnel for honesty. However, delivery persons have access to products (and even a truck is also available!) to remove as well as deliver goods. For this reason, it is important that receiving clerks work in an area with a clear view of delivery personnel and their vehicles.

Receiving areas must be kept extremely clean to avoid contaminating incoming food or providing a carrying vehicle for pests. Sometimes suppliers deliver goods that can harbor roach eggs or other insects. A clean receiving area makes it easier to both prevent and detect this type of problem. The area should be well-lit and properly ventilated. Too little light may cause product defects to go unnoticed, and excessive heat in the receiving area can quickly damage delivered goods, especially if they are refrigerated or frozen products.

Flooring in receiving areas should be light in color and of a type that is easily cleaned. In colder climates, it is important that the receiving area be warm enough to allow the receiving clerk to carefully inspect products. For example, if the outside temperature is below freezing, the outside loading dock is no place for an employee to conduct a thorough inspection of incoming products!

### Proper Tools and Equipment

Although the specific tools and equipment needed for effective receiving vary by type and size of operation, some items are required in any receiving area. These include:

✓ Scales: Scales should be of two types: those accurate to the fraction of a pound (for large items) and those accurate to the fraction of an ounce (for smaller items including pre-portioned meats). Scales should be calibrated (adjusted) regularly to ensure their accuracy.
✓ Wheeled equipment: These items, whether hand trucks or carts, should be available so goods can be moved quickly and efficiently to their proper storage areas.
✓ Box cutters: These tools, when properly maintained and used, allow the receiving clerk to quickly remove excess packaging, and this in turn helps to accurately verify the quality of delivered products. Of course, care must be taken when using this tool so proper training is essential to minimize any safety hazards.
✓ Thermometers: Foods must be delivered at their proper storage temperatures. Operators must establish the range of temperatures they deem acceptable for product delivery. For most operators, the temperature ranges are:

| Acceptable Temperature Range for Delivered Products | | |
| --- | --- | --- |
| Item | °F | °C |
| Frozen foods | 10°F or less | −12°C or less |
| Refrigerated foods | 30°F to 45°F | −1°C to 7°C |

✓ Calculator: Suppliers' calculations should always be checked, especially if the invoice has been prepared by hand. It is most helpful if the calculator has a physical tape (as on an adding machine) that can be used by the receiving clerk when needed. The calculator should also be available in case the original invoice is either increased or decreased in amount. The reasons why include incorrect supplier pricing and/or some items listed on the delivery invoice that may not have been delivered. In addition, invoice totals will change when all or a portion of the delivery was rejected because the items were of substandard quality or did not meet product specs.

### Proper Delivery Schedules

In an ideal world, a foodservice operator will agree to accept delivery of products only during the hours they select. These times would likely be during the operation's slowest periods when there is plenty of time for a thorough check of the products being delivered.

In fact, some operators can demand that deliveries be made only at certain times, for example, between 9:00 a.m. and 11:00 a.m. These are called **acceptance hours,** and the operation may refuse to accept delivery of products at any other time.

**Key Term**

**Acceptance hours:** The hours of the day in which an operation will accept food and beverage deliveries.

Other operations prefer to establish times at which they will *not* accept deliveries, for example, between 11:00 a.m. and 1:00 p.m. when the operation is busy serving lunch. These are called **refusal hours.** A busy lunchtime may make it inconvenient to accept deliveries, and some operators will simply not take deliveries then. In both

**Key Term**

**Refusal hours:** The hours of the day in which an operation will not accept food and beverage deliveries.

cases, however, the assumption is that the operator is either a large enough or a good enough customer to make demands such as these. If an operation does establish either acceptance or refusal hours, these should be clearly communicated to and agreed to by the operation's suppliers.

In many operations, and especially smaller ones, it is the supplier who will determine when products are to be delivered. This may seem inconvenient (and often it is!), but it is important to recognize that all foodservice operators would prefer to have deliveries made during their slow periods (between their peak mealtimes). In many cases, it is simply not possible for suppliers to stop their trucks for several hours to wait for a preferred delivery time. In fact, some foodservice operators in remote locations will only be told the day a delivery will be made rather than a specific time of day.

The key to establishing a successful delivery schedule with suppliers is, quite simply, to communicate with them. While every relationship between operator and supplier may differ, when both sides are working together, they can generally come to an acceptable agreement about the timing of product deliveries.

## Key Areas of Product Verification

All receiving personnel must be familiar with their operation's product specs, and they must be properly trained. When they are well-trained, they can readily provide necessary inspections to verify the following key characteristics of delivered products:

1) Weight
2) Quantity
3) Quality
4) Price

### Weight

In many cases, one of the most important items to verify when receiving food products is their weight. However, doing this is not always easy. For example, a 14-pound package of ground beef will look exactly like a 15-pound package, and there is no easy way to tell the difference without putting the product on an accurate scale. Receiving clerks should be required to weigh all incoming meat, fish, and poultry. Note: One exception is unopened Cryovac (sealed) packages containing items such as hot dogs and bacon. In this situation, the entire case should be weighed to detect shortages in content.

Often, meat deliveries will consist of several items, all of which are packaged together in one box or container. When the receiving clerk is very busy, the temptation exists to just weigh the entire box of all products together. To illustrate why this must not be done, consider the case of Lea Robbins.

Lea's operation serves a variety of menu items, the production of which requires ground beef, New York Strip steaks, and corned beef. Assume Lea ordered 40 pounds of product from Bart's Meats. The portion of the PO detailing the items Lea ordered is shown in Figure 4.2.

When the Bart's Meats delivery person arrived, all three items were packed into one cardboard box, and the delivery person was in a hurry. He, therefore, suggested that Lea simply weigh the entire box. When she did, Lea discovered that

| Item Ordered | Unit Price | Total Ordered | Total Price |
|---|---|---|---|
| Ground Beef | $ 4.50/lb. | 10 lb. | $ 45.00 |
| New York Strip Steak | 14.00/lb. | 20 lb. | 280.00 |
| Corned Beef | 18.60/lb. | 10 lb. | 186.00 |
| Total | | 40 lb. | $ 511.00 |

**Figure 4.2**  Lea's Ordered Items

| Item Ordered | Unit Price | Total Weight Delivered | Actual Value |
|---|---|---|---|
| Ground Beef | $ 4.50/lb. | 15 lb. | $ 67.50 |
| New York Strip Steak | 14.00/lb. | 10 lb. | 140.00 |
| Corned Beef | 18.60/lb. | 15 lb. | 279.00 |
| Total | | 40 lb. | $486.50 |

**Figure 4.3**  Lea's Delivered Items

the contents weighed 40.5 pounds. Since the box itself weighed about ½ pound, she signed for the delivery, agreeing that $511 worth of product had been received. When she began to put the meat away, however, she did weigh each item individually and found the information in Figure 4.3.

If Lea called her supplier to complain about the overcharge ($511 total price as per the PO −$486.50 actual value of delivered items = $24.50 overcharge), she would likely be told that the delivery problem she noticed was simply a mistake caused by human error. It may well have been, but the lesson here is that, when an item is ordered by weight, its delivery must be verified by weight. It is up to the operator to train receiving clerks to always verify that the operation is charged only for the product weight delivered. Excess packaging, ice, or water in the case of produce can all serve to increase the delivered weight. The effective receiving clerk must be aware of and be on guard against deceptive delivery practices.

## Quantity

The proper counting of products is often as important as proper weighing. Products delivered but not charged for cost the supplier money. Products not delivered but charged for cost the operation money and more.

For example, if an operator orders three cases of whiskey, the operator wants to receive and pay for three cases of the product. This is important for two reasons. First, the operator only wants to pay for products that have been delivered. Second, and just as important, if the buyer prepared the PO correctly, then three cases of whiskey were needed. If only two cases are delivered, the operation may not be able to prepare enough of the whiskey-based drinks its guests want to order. If this means the operation will run out of the product or make a substitute, it may result in unhappy guests.

**Short** is the term used in the foodservice industry to indicate that an ordered item has not been delivered. When a supplier shorts the delivery of an ordered item, that item may or may not appear on the delivery invoice. If it does not appear, it must be noted so that management knows that the item is missing, and appropriate reorder action can be taken.

**Key Term**

**Short (delivery item):** When a supplier does not deliver the quantity of an item ordered for the appointed delivery date.

If the undelivered item is, in fact, listed on the delivery invoice, the delivery driver should sign a **credit memo**. A credit memo indicates an adjustment to a delivery invoice must be made. In most cases, credit memos should be prepared in triplicate (three copies). One copy is retained in the receiving area to be filed. One copy is retained by management for proper bill payment, and the original copy is given to the supplier.

**Key Term**

**Credit memo:** An addendum to a supplier's delivery invoice that reconciles differences between a delivery invoice and the product(s) received.

Figure 4.4 is an example of a credit memo. Note that the credit memo has a place for the signature of a representative from the operation and the supplier. Both must sign it.

---

**CREDIT MEMO**

**Name of Operation** _____

**Vendor** _____     **Delivery Date** _____

**Invoice #** _____     **Credit Memo #** _____

| Correction(s) | | | | | |
|---|---|---|---|---|---|
| Item | Quantity | Short | Refused | Unit Price | Total Amount |
|  |  |  |  |  |  |
|  |  |  |  |  |  |
|  |  |  |  |  |  |
|  |  |  |  |  |  |
|  |  |  |  |  |  |
| Total |  |  |  |  |  |

Original Invoice Total _____

Less: Credit Memo Total _____

Adjusted Invoice Total _____

Additional Notes: _____

_____

Vendor Representative _____

Vendor Representative Contact Number _____

Operation Representative _____

Operation Representative Contact Number_____

---

**Figure 4.4** Sample Credit Memo

The credit memo is a formal way of notifying a supplier that an item listed on the original delivery invoice is missing or defective, and, therefore, the value of that item should be deducted from the invoice's total. If a supplier consistently shorts an operation, that supplier is suspect in terms of both honesty and lack of concern for the operation's long-term success.

---

**Technology at Work**

Foodservice operators use credit memos to record adjustments to delivery invoices. Credit memos are typically used to decrease the amount that is owed. In larger foodservice operations or those that have large numbers of deliveries, the initiation of credit memos and their monitoring and management can become quite complex.

Fortunately, several companies offer software programs specifically designed to allow business owners to apply credit memos against the correct invoices to ensure that the resulting invoice amounts that must be paid are correct. These programs allow operators to enter credit memo dates, credit amounts before and after tax, credit memo identification numbers, and descriptions of the reasons why a credit memo is being generated.

To learn more about stand-alone software that can be used to manage credit memos, enter "credit memo software for restaurants" in your favorite search engine and view the results.

---

The counting of boxes, cases, bags, and other containers must be routine behavior for the receiving clerk. Counting items such as the number of lemons or oranges in a box should be done on a periodic basis, but the value of counting items such as these on a regular basis is questionable. Although an unscrupulous supplier might be able to remove one or two lemons from each box delivered, the time required to detect such behavior is often not worth the effort required to spot it. It is preferable to do a thorough item count of such items only periodically and to work only with reputable suppliers.

The direct delivery of products to a foodservice operator's storeroom or holding area is another area for concern. The delivery person may deliver some items including bread, milk, and carbonated beverage mixes directly to their respective storage areas but doing so often bypasses the receiving clerk. This should not be allowed because, when this action is taken, it may be impossible to verify the accurate quantity of items delivered. If this process must be used, product dates on each item can help ensure that all products listed on the delivery invoice were received.

## Quality

No area of an operation should be of greater concern to managers than the delivery of the appropriate quality of products. If an operation takes the time to develop product specs, but then accepts delivery of products that do not match these specs, both time and effort are wasted. Without product specs, verification of quality is difficult because the receiving and management staff will be unsure of the quality level that is desired. Suppliers know their products, and they also know their customers. Some operators will only accept the quality of what they have specified. Others will accept virtually anything because they do not inspect or verify deliveries. If you were a supplier, and you had a sack of onions that was getting past its prime, this product would likely be delivered to the operator that does not rotate products.

Specs are also necessary because they "even the playing field," and all potential suppliers must quote prices for items meeting the spec's quality standards. Wise operators are aware that less ethical vendors may quote relatively low prices for items that meet an operation's specs, and they then substitute products that do not meet the specs. When (or if) operators inform these suppliers about these problems, they can simply exchange items "without additional charge." Note: At the same time, they learn that the operator does check specs and, in the future, must be quoted a price that represents an item of the proper quality.

Unscrupulous or careless suppliers can cost an operation both in price paid and quality received. Consider, for example, the foodservice operator who specifies that no more than a ¼-inch fat cover on all New York strip steaks ordered is acceptable for this product. Instead, the meat company delivers steaks with a ½-inch fat cover. The operation will, in this example, pay too much for the product because steaks with a ¼-inch fat covering are sold by suppliers at a higher price per pound than those with a ½-inch covering. When the steaks are served, guests will likely be less happy with a steak that looks too "fat" than they will be with one that looks just right!

Checking for quality means checking the entire shipment for conformance to established product specs. If only the top row of tomatoes in the box conforms to the operation's spec, it is up to the receiving clerk or manager to point that out to the suppliers. If the remaining tomatoes in the box do not meet the spec, the box should be refused. A credit memo can then be used to reduce the total on the invoice to the proper amount.

In some cases, quality deficiencies are not discovered until after a delivery driver has left an operation. Then the operator should immediately notify the supplier that a thorough inspection has uncovered some substandard products. The supplier may then be instructed to pick up the nonconforming items. Many operators use the "Additional Information" section of the credit memo form shown in

Figure 4.4 to record this requested pickup. When the product is picked up, the required information is then recorded.

Alternatively, a separate memo/e-mail/text to the supplier requesting a product pickup could be generated. It is best, however, to keep the number of cost control forms to a minimum whenever possible, especially when minor modifications of one form will allow that form to serve two purposes.

---

**Find Out More**

In many foodservice operations, the identification of high-quality produce is one of a receiving clerk's most important skills. One good step in becoming knowledgeable about produce standards and quality is obtaining a copy of the *Fresh Produce Manual* published by the Produce Marketing Association (PMA).

Founded in 1949, PMA is a not-for-profit global trade association serving its members who market fresh fruits, vegetables, and related products worldwide. Its members are involved in the production, distribution, retail, and foodservice sectors of the industry. PMA offers its members a wide variety of services, including the publication of many materials related to fresh produce quality and standards.

PMA's *Fresh Produce Manual* is a full-color guide to the identification of high-quality fruits and vegetables. It addresses such characteristics as varieties sold, grades, sizes, storing, handling, and common packaging for a wide range of products, and it can be an invaluable training aid for new receiving clerks.

To see an online version of the *Fresh Produce Manual*, enter "www .americanfruitandproduce.com/uploads/data/pm.pdf" in your favorite search engine and view the results.

---

**What Would You Do? 4.1**

"This looks pretty bad," said Chef Allen as he carefully examined a head of iceberg lettuce taken from the 24-head case that he was being shown by his receiving clerk.

"The whole case looks like that," said Shingi, the Chef's receiving clerk. Later, when Chef Allen called his produce sales representative, he found out why.

"I know it looks kind of "rusty" in spots and the trim loss rate will be a lot higher," said Lisa, the salesperson for the produce supplier.

"Well why did you send me rusty lettuce?" asked Chef Allen.

"Actually Chef, it's called 'Downy Mildew,' and it happens when there is too much rain to harvest the lettuce heads at the right time. My own suppliers tell me it won't improve for several more days. I can't change that, but we do have

*(Continued)*

other unaffected greens in stock. Right now, I can send you butterhead, Bibb, looseleaf, romaine, watercress, arugula, oak leaf, escarole, or radicchio. Would you like to place a re-order?"

**Assume you were Chef Allen. If the product specification for your salads called for iceberg lettuce, what are the advantages to using the iceberg lettuce and simply absorbing the higher trim loss? What could be the disadvantages associated with keeping the iceberg lettuce and absorbing the higher trim loss? Why would knowledge about alternative salad greens and their characteristics be helpful to you and your receiving clerk in this situation?**

Whether the item purchased is a food item or an alcoholic beverage, suppliers sometimes run out of a product, just as an operator may sometimes run out of a menu item. In such cases, the receiving clerk must know whether it is management's preference to accept a product of higher quality, lower quality, or no product at all as a substitute. If this information is not known, one can expect that suppliers will be able to downgrade quality simply by saying that they were "out" of the requested product and did not want the operator to be "shorted" on the delivery.

Training to assess and evaluate quality products is a continuous process. The effective receiving clerk should develop a keen eye for quality. It is important that both the operation and the guests it serves receive the quality of products intended to be served.

---

**Find Out More**

Foodservice operators purchasing fresh seafood face some unique challenges as they inspect and receive requested products. Simply put, much of the seafood sold in the United States (and other parts of the world) is mislabeled.

In fact, in a study summary released by the UK-based *The Guardian* newspaper in 2021, it was found that "36 percent of over 9,000 (seafood) products were mislabeled. And the issue isn't just afflicting other parts of the world: Of the countries included, the United States was the third worst, with 38 percent of seafood mislabeled—behind only the United Kingdom and Canada (both of which had an appalling 55 percent mislabeling rate) The problem goes beyond ill-informed grocery shoppers, as well. One study published in 2018 collected samples from 180 restaurants across 23 European countries and found that a third of those restaurants had sold mislabeled seafood—with the odds of some menu items being authentic equal to a coin toss."

A 2019 CNN report on the same issue found that "Favorites like sea bass and snapper had some of the highest rates of mislabeling. Sea bass was mislabeled

55% of the time and snapper 42% of the time, Oceana's tests showed. Often, instead of sea bass, they'd get giant perch or Nile tilapia, fish that should be less expensive and is considered lower quality. Dover sole they tested was actually walleye. Lavendar jobfish had been substituted for Florida snapper."**

While some amount of mislabeling of seafood is likely caused by innocent supplier error, foodservice operators must be extremely careful when purchasing fresh seafood. To find out more about this continuing industry problem, enter "mislabeled seafood in the restaurant industry" in your favorite search engine and view the results.

*www.foodandwine.com/news/seafood-fraud-rampant-new-report, retrieved January 25, 2023.

**www.cnn.com/2019/03/07/health/fish-mislabeling-investigation-oceana/index.html, retrieved January 25, 2023.

## Price

When receiving staff verify price during delivery, three major concerns must be addressed:

✓ Matching PO unit price to delivery invoice unit price
✓ Verifying price extensions
✓ Verifying invoice totals

### Matching PO Unit Price to Delivery Invoice Unit Price

A product's pricing unit (see Chapter 3) may be pounds, quarts, gallons, cases, or other commonly used measurement unit for which a price has been established. An item's **purchase unit,** however, may be different than its pricing unit.

For example, bacon may be priced by the pound, but purchased in 3-, 5-, or 10-pound packages. Similarly, eggs may be priced by the dozen, but sold in cases containing six trays holding 30

> **Key Term**
>
> **Purchase unit:** The packaging configuration (e.g., case, carton, bushel, box, or bag) by which a supplier would normally sell a product.

eggs each (180 eggs). In many cases, frozen meats, poultry, and fish will be priced by the pound or kilo, but they are sold in containers of various sizes. Regardless of the purchase price or purchase unit, receiving clerks must assure that the prices listed on the delivery invoice are the same as the prices listed on the associated PO.

When the person responsible for purchasing food for the operation places a PO, the confirmed quoted price should be carefully recorded because it is never safe to assume that the delivered price will match the price listed on the PO. Ethical foodservice operators should not be happy with either an overcharge or an undercharge for a purchased product. Just as operators would hope that a guest would inform them if a service staff member forgot to add the correct price of a bottle of wine to a dinner check, ethical receiving clerks work with their suppliers to ensure that the operation is neither under- nor over-charged for all items delivered. Honesty and fair play must govern the actions of both the operator and the supplier.

If the receiving clerk has access to the original PO, it is a simple matter to verify the quoted price and the delivered price. If these numbers do not match, management should be notified immediately. If management notification is not possible, both the driver and the receiving clerk should initial the "Comments" section of the PO, showing the difference in the two prices, and a credit memo should be prepared.

It is important to recognize that, if the receiving clerk has no record of the initially quoted purchase price from either a PO or an equivalent source, price verification of this type is not possible. An inability to verify the quoted price of an item and its delivered price at the time of acceptance is a sure indication of a poorly designed receiving system.

Some operators deal with suppliers in a way that establishes a **contract price**. A contract price arrangement is simply an agreement between the buyer and seller to hold the price of a product constant over a defined period of time. For example, assume an operator uses Dairy-O'Fresh as a milk supplier.

Dairy O'Fresh agrees to supply the operator with milk at the price of $4.55 per gallon from January 1 through March 31 this year. The operator is free to buy as much or as little milk as needed. The milk will always be billed at the contract price of $4.55 per gallon.

### Key Term

**Contract price:** A price mutually agreed upon by supplier and operator. This price is the amount to be paid for a product or products over a prescribed period of time.

The advantage to the operators of this arrangement is that they know exactly what the per-gallon milk cost will be for the 3-month period. The advantage to Dairy O'Fresh is that it can offer a lower price in the hope of securing all of the operator's milk business. Even in the case of a contract price, however, the receiving clerk should verify the invoice delivery price against the established contract price.

### Verifying Price Extensions and Total

Verification of price extension is just as important for operators to monitor as is the ordered and delivered price. Price extension is the process by which operators compute an **extended price**.

Extended price is simply the purchase unit price multiplied by the number of units delivered.

### Key Term

**Extended price:** The price per purchase unit multiplied by the number of units delivered. This refers to the total per product price as listed on a delivery invoice.

For example, if the price for a case of lettuce is $44.55 and the number of cases delivered is six, the extended price is calculated as:

Purchase unit price × Number of units delivered = Extended price

Or

$44.55 per case × 6 cases delivered = $267.30

Extended price verification is extremely important. It is critical that the receiving clerk verify:

✓ Purchase unit price
✓ Number of units delivered
✓ Extended price computations (purchase unit price × number of units delivered)
✓ Invoice totals

There may be two major reasons why operators do not always insist that the receiving clerk verify extended prices and invoice totals. The most common reason is the belief that there is not enough time to do so. The driver may be in a hurry, and the operation may be very busy. If that is the case, the process of verifying the extended price can be moved to a slower time. Why? There is a written record provided by the supplier of both the purchase unit price and the number of units delivered. Extension errors become supplier errors that are recognized by the supplier in the supplier's own handwriting! Or, more accurately, today they are in the supplier's own computer system.

The second reason operators sometimes ignore extended price verification is related to these same computers. Some operators believe that if an invoice is machine-generated, the mathematics of price extension must be correct. Nothing could be further from the truth. Anyone familiar with the process of using computers knows that there are many possible entry errors that can result in extension errors (even formulas entered onto Excel-type spreadsheets can be entered in error). Once all extension prices have been verified as correct, receiving clerks should check the invoice total against the sum of the individual price extensions. Managers must ensure that delivery clerks verify both extended prices and invoice totals. If this cannot be done at the time of delivery, it must be done as soon thereafter as is reasonably possible.

## Receiving Records

The records generated during the receiving process are important and must be properly maintained if an operation's **accounts payable (AP)** amounts are to be recorded accurately.

### Key Term

**Accounts payable (AP):** Money owed by the foodservice operation to suppliers or others that has not yet been paid. Most often referred to as "AP."

Suppliers doing business with a foodservice operation produce a written record of the amount the supplier believes is owed to them for products that have been delivered. Similarly, the foodservice operation will have a record of what they believe is owed to the supplier. In many cases these records will match, but in some cases they will not. For this reason, it is important that a foodservice operation maintain receiving records generated by their suppliers and those generated by the operation.

## Supplier-Produced Delivery Records

When a supplier is relatively small, their own delivery records may simply consist of the hard copy delivery invoices generated for their foodservice operation customers and given to the operation at the time of product delivery. In the case of larger suppliers, and in increasing numbers with smaller suppliers, delivery records are maintained online in an operation's designated client account.

The importance of an operation retaining supplier delivery records is easy to understand when supplier invoicing procedures are considered. For example, assume a bread company makes deliveries three times a week to a foodservice operation. On each delivery date, the supplier provides a delivery invoice detailing the products that have been delivered that day.

The supplier, however, will not likely invoice the foodservice operation three times a week. Rather, the supplier will periodically invoice the operator. The invoice may be weekly, bi-weekly, or even monthly. In this scenario, a foodservice operator paying the bread supplier must have the detail of each day's delivery if they are to compare the sum of each single day's deliveries to the amount listed on the supplier's combined (multi-delivery day) invoice.

If a foodservice operation does not keep a record of the individual delivery invoices produced by the supplier, they will not be able to match the delivery invoice numbers with those listed on the periodic invoices received from the supplier. Maintaining a copy of the supplier-produced delivery records helps ensure the individual responsible for an operation's AP does not pay for the same delivery invoice more than one time.

### Operation-Produced Delivery Records

Some larger foodservice operations produce a daily receiving record when accepting delivery of food and beverage products. This method, which takes extra administrative time to both prepare and monitor, has some recordkeeping advantages for both large and small operations.

A daily receiving record will generally contain the following information:

1) Date
2) Name of receiving clerk verifying deliveries
3) Name of supplier

4) Supplier's delivery invoice number
5) Total delivery invoice amount
6) Purchase order (PO) number
7) Purchase order (PO) amount
8) Variance (if any)
9) Notes/comments section

Figure 4.5 is an example of a receiving record produced by Sally Rodgers, the owner/operator of Sally's Open Pit Bar BBQ.

---

**Sally's Open Pit BBQ**

**RECEIVING RECORD**

Date _____                     Receiving Clerk _____

| Supplier | Supplier Invoice # | Delivery Invoice Amount | PO # | PO Amount | Variance |
|----------|--------------------|--------------------------|------|-----------|----------|
|          |                    |                          |      |           |          |
|          |                    |                          |      |           |          |
|          |                    |                          |      |           |          |
|          |                    |                          |      |           |          |
|          |                    |                          |      |           |          |
|          |                    |                          |      |           |          |
|          |                    |                          |      |           |          |
|          |                    |                          |      |           |          |
|          |                    |                          |      |           |          |
|          |                    |                          |      |           |          |
|          |                    |                          |      |           |          |
|          |                    |                          |      |           |          |
|          |                    |                          |      |           |          |
|          |                    |                          |      |           |          |

Notes and comments _____
_____
_____
_____
_____
_____

**Figure 4.5**  Sample Receiving Record

Note that when using a receiving record foodservice operators can match delivery invoices against original POs, identify any variation between the two, and include comments about actions taken (e.g., the issuing of a credit memo or the reorder of shorted products).

Regardless of whether they rely only on supplier-provided records of deliveries or create their own records, every foodservice operation must have a system in place to document the receipt of ordered products. If they do not, they will be susceptible to errors made in AP and to potential **billing scams**.

While billing scams can take a variety of forms, the three types most often encountered by foodservice operators are:

**Key Term**

**Billing scam:** An effort to trick an operation into paying for products or services that it did not order, that have little or no value, or that were never delivered. Also known as an "invoice scam."

1) Invoices for unordered products

In this scam, an invoice arrives for products that were delivered but not ordered. In many cases, the invoice amounts will be much greater than the value of the delivered products. Even if an employee in a foodservice operation has signed for and accepted the products, it is important to recognize that there is no obligation to pay for them. Nor are operators required to return the products to the vendor. Under most state laws, the receipt of unsolicited goods is considered a gift, and the gift recipient may use or dispose of the items as they see fit.

2) Fictitious billings

In a fictitious billing scam, a fraudster creates a fictitious vendor/supplier who then bills the foodservice operation for payment. In this scam, the fraudster creates a fake supplier name and post office box for receipt of payments, which are then kept by the individual(s) sending the fictitious billing.

3) Duplicate invoice payments

In this scam, the fraudster utilizes the name of a legitimate supplier to try to cause double payment. This scam may be difficult to detect because the individual in charge of an operation's AP will likely recognize the supplier's name and assume the invoice is legitimate. The best defense against this scam is to have a system in place that allows invoices to be matched to an operation's internally created delivery records.

Foodservice operators can take specific steps to avoid billing scams:

✓ Limit the number of individuals who are authorized to order products and supplies.
✓ Closely monitor all invoices and especially those from unfamiliar suppliers.

✓ Pay particular attention to ensuring that the operation's account number on the invoice matches the account number provided by the supplier. If the numbers are not the same, it is likely a billing scam.

✓ Pay particular attention to any invoice that does not have a telephone number to contact the supplier, as those perpetuating a billing scam will not likely provide such contact information.

✓ Prior to payment, confirm all requests for payment with the individual who authorized the initial purchase.

✓ Do not pay invoices unless it is confirmed that products listed on the invoice were ordered and delivered.

---

**Technology at Work**

Regardless of the accounting software system utilized in their businesses, foodservice operators often find that the management of accounts payable (AP) can be extremely time intensive and sometimes very frustrating.

Foodservice operators managing their AP face a variety of challenges. The first concerns too much paper. From paper delivery invoices to paper checks, a reliance on hard copy documentation slows down the AP process. Converting as much of the AP process as possible to electronic records speeds up an operation's ability to store and retrieve critical payment information. An additional AP challenge relates to the chance for simple human error. Unless a foodservice operator's AP records are meticulously maintained, the chance for simple mistakes resulting in duplicate payments or incorrect amounts paid can be common.

One of the most important aspects of AP management relates to the timing of payments. Keeping track of various supplier's payment terms can be an arduous task. If supplier bills are paid too early, it can result in possible operating cash shortages. Bills paid too late, however, can sometimes result in late payment penalties. As a result, deciding which suppliers to pay and when to pay them are important concerns in the AP management process.

Fortunately, several companies produce foodservice-specific accounts payable software that can assist with these AP-related issues and more. To learn more about these useful programs, enter "best accounts payable software for restaurants" in your favorite search engine and review the results.

The proper receiving and documenting of ordered products is an important cost control step. The proper storage and inventory of an operation's products after they are received and documented is another important step, and that is the topic of the next chapter.

---

**What Would You Do? 4.2**

"Come on Nicole, give me a break. I'm already behind because my truck broke down this morning, and I've got people all over town calling my boss to scream about their deliveries. It wasn't my fault I'm late. I'm just the driver," said Mike.

Mike, who makes deliveries for Wolverine Produce, was talking to Nicole, the kitchen manager and receiving clerk for the High-Five Restaurant.

The delivery Mike was making was a big one, and it was two hours late. "You know, Nicole," Mike continued, "you folks take longer to accept a delivery than any other restaurant on my route. Nobody else inspects and weighs like you do. And you hardly ever find any problems."

Mike paused, then said "Look, I know it's a big delivery, but just this once can't you just sign the invoice and let me get going? I want to see my daughter's basketball game, and I won't make it if you take forever to inspect this load."

"I don't know Mike," replied Nicole. "We've got procedures to follow here, and I'm supposed to use them every time."

"Just, sign the ticket. If you find a problem later, I'll take care of it. I promise you," said Mike, who appeared to be increasingly flustered.

**Assume you were Nicole. How would you likely respond to Mike's request that you speed up? Now assume you were this restaurant's owner. If you were personally accepting the delivery, how would you likely respond to Mike's request?**

## Key Terms

| | | |
|---|---|---|
| Auditor | Short (delivery item) | Extended price |
| Delivery invoice | Credit memo | Accounts payable (AP) |
| Acceptance hours | Purchase unit | Billing scam |
| Refusal hours | Contract price | |

## Operator's 10-Point Tactics for Success Checklist

Evaluate your need for, and the current status of, each of the following operational tactics. For those tactics you think are important, but not yet in place, develop an action plan for its implementation including who will be responsible for the tactic's completion and the target date by which it should be completed.

| Tactic | Don't Agree (Not Done) | Agree (Done) | Agree (Not Done) | If Not Done | |
|---|---|---|---|---|---|
| | | | | Who Is Responsible? | Target Completion Date |
| 1) Operator can identify the various steps that must be completed when properly receiving product deliveries. | —— | —— | —— | | |
| 2) Operator recognizes the importance of proper location when developing a professional receiving program. | —— | —— | —— | | |
| 3) Operator recognizes the importance of proper tools and equipment when developing a professional receiving program. | —— | —— | —— | | |
| 4) Operator recognizes the importance of proper delivery schedules when developing a professional receiving program. | —— | —— | —— | | |
| 5) Operator has adequately trained receiving staff about how to verify the weight of delivered products. | —— | —— | —— | | |
| 6) Operator has adequately trained receiving staff about how to verify the quantity of delivered products. | —— | —— | —— | | |
| 7) Operator has adequately trained receiving staff about how to verify the quality of delivered products. | —— | —— | —— | | |

*(Continued)*

| Tactic | Don't Agree (Not Done) | Agree (Done) | Agree (Not Done) | If Not Done | |
|---|---|---|---|---|---|
| | | | | Who Is Responsible? | Target Completion Date |
| 8) Operator has adequately trained receiving staff about how to verify the price of delivered products. | —— | —— | —— | | |
| 9) Operator understands the importance of retaining supplier-initiated delivery records. | —— | —— | —— | | |
| 10) Operator understands the importance of creating and retaining operation-initiated product delivery records. | —— | —— | —— | | |

# 5

# Cost Control in Storage and Issuing

**What You Will Learn**

1) The Importance of Cost Control in Storage and Issuing
2) The Essentials of Professional Storage
3) How to Determine the Value of Product Inventories
4) How to Issue and Restock Product Inventories

**Operator's Brief**

In this chapter, you will learn that controlling costs during product storage, issuing, and restocking is essential to your foodservice operation's profitability.

The two concerns of primary importance when storing inventory items are the maintenance of quality and security. If inventory items are not properly stored, the result can be loss of product quality and/or quantity. Whether they are stored in dry-storage areas, refrigerators, or freezers, you must take care to ensure that all food items in inventory are held in ways that help maintain their optimum quality levels.

In addition, in this chapter, you will learn that product inventories must be properly secured against the potential for theft if storage costs are to be effectively managed.

The proper calculation of the monetary value of the products you hold in your inventory is essential for calculating product usage costs, and in this chapter, you will learn how it is done. You will also learn that the proper issuing and restocking of product inventories is critical, and how the use of a physical, perpetual, or ABC inventory control system can help focus your attention on those stored items of most importance to your operation's success.

*(Continued)*

Finally, in this chapter, you will learn that some inventoried items are best purchased (restocked) utilizing an "as-needed" approach, while others are best purchased using a "par-level" approach. Both approaches are described in detail and understanding them will allow you to establish purchase (restocking) points that makes good sense for each of your stored inventory items.

---

**CHAPTER OUTLINE**

The Importance of Cost Control in Storage and Issuing
Storage Essentials
    Quality Concerns
    Security Concerns
Valuing Product Inventories
Issuing and Restocking Product Inventory
    Issuing
    ABC Inventory Control
    Restocking Product Inventories

---

## The Importance of Cost Control in Storage and Issuing

The previous two chapters presented the procedures used to properly purchase and receive foodservice products. In most cases, after purchased items have been received, they must be immediately placed into storage.

Food and beverage product quality rarely improves with increased storage time. In fact, the quality of most products foodservice operators buy will be at their peak levels when the products ordered are initially delivered. From then on, many food and beverage products will decline in freshness, quality, and nutritional value.

It is important to understand that all stored food and beverage products that make up an operation's **product inventory** should be thought of in their monetary value.

For example, an 8-ounce New York strip steak in a walk-in is not just a steak. It represents, at a selling price of $20.00, that amount of revenue.

**Key Term**

**Product inventory:** All of the stored food and beverage items used to produce an operation's menu items.

If the steak disappears, revenue of $20.00 will disappear also. When operators think of their product inventory in terms of their sales value, it becomes clear why food and beverage products must be stored and maintained carefully.

## Storage Essentials

The ideal situation for foodservice operators would be to store only the food and beverage products they will use between the time of a supplier's delivery and the time of that supplier's next delivery. This is true because storage costs money in terms of providing both the storage space for inventory items and the money tied up in inventory items and unavailable for use elsewhere.

When possible, it is best to order only the products that are absolutely needed by an operation. In that way, the supplier's storeroom actually becomes the operation's storeroom because the supplier then absorbs the costs of storing needed products. In all cases, however, an operation must have an adequate supply of products on hand to serve its guests. Guests should be an operation's main concern. If an operation is popular, it will have many guests and will need many items in storage!

After receiving staff have properly accepted the food products that were purchased, the next step in the control of product costs is that of properly storing those items. In most cases, the storage process consists of two main concerns:

1) Quality concerns
2) Security concerns

### Quality Concerns

Many food products are highly perishable items, and they must be moved quickly from an operation's receiving area to the area selected for storage. This is especially true for refrigerated and frozen items. An item such as ice cream, for example, can deteriorate substantially in quality if it remains at room temperature for only a few minutes. Most often in foodservice, this high perishability means that the same individual responsible for receiving the items is the one who is responsible for their storage.

Consider the situation of Kendra, the receiving clerk at Fairview Estates, an extended care facility with 200 residents. She has just taken delivery of seven loaves of bread. They were delivered in accordance with the purchase order her manager prepared. The bread must now be put in storage. When Kendra stores these items, she must know whether management prefers her to use the **LIFO (last-in first-out)** or the **FIFO (first-in first-out)** method of product rotation.

**Key Term**

**LIFO (last-in first-out) inventory system:** An inventory management system in which the most recently delivered products are used before products already in storage.

**Key Term**

**FIFO (first-in first-out) inventory system:** An inventory management system in which products already in storage are used before more recently delivered products.

## LIFO System

When using the LIFO storage system, an operation intends to use the most recently delivered product (last-in) before it uses any part of that same product previously on hand. If Kendra is instructed, for example, to use the just-delivered bread *prior* to using other bread already in storage, she would be utilizing the LIFO system.

In all cases, operators must strive to maintain a consistent product standard. In the case of some selected bread, and some fresh pastry items, the receiving clerk could be instructed to utilize the LIFO system. With LIFO, great care must be taken to order only the quantity of product needed between deliveries. If too much product is ordered, loss rates may be high.

For most of the items a foodservice operation will buy, however, the best storage system to use is the FIFO storage system.

## FIFO System

With a FIFO system, older products (first-in) are used before newer products. When the FIFO system is used, the storeroom clerk must take great care to place new stock behind, underneath, or at the bottom of old stock. It is often the tendency of employees not to do this.

Consider, for example, the storeroom clerk who must put away six cases of tomato sauce. The cases weigh about 40 pounds each. The FIFO method dictates that these six cases be placed *under* the five cases already stacked in the storeroom. Will the clerk take the time and effort to place the six newly delivered cases underneath the five older cases? In many instances, the answer is "no."

Unless management strictly enforces the FIFO rule, employees may be tempted to use the faster and easier, but improper, way of storing the newly delivered products *on top* of those already in storage. Figure 5.1 shows the difference between LIFO and FIFO when dealing with storing boxes, bags, or cases of food products, with items at the top of the list being stored on top so they will be used first.

| Product Inventory Storage | |
| --- | --- |
| Oldest | Newest |
| Oldest | Newest |
| Oldest | Newest |
| Newest | Oldest |
| Newest | Oldest |
| Newest | Oldest |
| **FIFO** | **LIFO** |

**Figure 5.1** Stacking Items with FIFO and LIFO Storage Systems

FIFO is the preferred storage technique for most perishable and non-perishable items. Failure to implement a FIFO system of storage management can result in excessive product loss due to spoilage, shrinkage, or deterioration of product quality.

Decisions about storing food items according to the LIFO or FIFO method are management decisions. Once these decisions have been made, they must be communicated to storeroom staff and monitored on a regular basis to ensure compliance.

To assist them in their efforts, some foodservice operators require the receiving or storeroom clerk to mark or tag each delivered item with the date of delivery. These markings provide a visual aid in determining which products should be used first. This is especially critical in the area of highly perishable and greater cost items such as produce, fresh meats, and seafood.

Special meat and seafood date tags are typically available from meat and seafood suppliers. These tags contain a space for writing in the item's name, quantity, and delivery date. The use of these tags or an alternative date tracking system is strongly recommended. If a supplier has computerized their delivery system, the box or case delivered may already bear a bar code strip or **QR code** identifying both the product and the delivery date. When this is not the case, however, the storeroom clerk should perform this function.

### Key Term

**QR code:** A QR (quick response) code is a machine-readable bar code that, when read by the proper smart device, allows inventory items to be identified by name, price, and delivery date.

### Find Out More

All foodservice operators must be interested in serving foods at peak quality levels. Knowing when a food product has reached, or exceeded, its peak quality level, however, can be challenging. It's important to recognize that, except for infant formula, product dating of foods is not required by federal regulations. Therefore, there are no uniform or universally accepted descriptions used on food labels for open dating in the United States. As a result, there are a wide variety of phrases food manufacturers and suppliers voluntarily use on labels to describe quality dates.

Examples of commonly used phrases include*:

✓ A "Best-if-Used-By/Before" date that indicates when a product will be of best flavor or quality.
✓ A "Sell-By" date tells a store how long to display the product for sale. It is not a safety date.

*(Continued)*

> ✓ A "Use-By" date that is the last date recommended for use of the product while at its peak quality. This is not a safety date (except for when used on infant formula), and it does not indicate an item is unsafe or must be disposed of after this date.
>
> ✓ A "Freeze-By" date indicates when a product should be frozen to maintain peak quality.
>
> In many cases, foodservice operators will create their own "Use-By" dates. These may be dates, for example, which indicate how long a pre-prepared food can be stored in the refrigerator or freezer. Additional examples include the dating of leftovers to indicate the dates by which they must be used, and the dating of menu items made with a variety of ingredients, each of which may have different "Best-if-Used-By" dates.
>
> Those foodservice operators labeling stored foods in-house will find a variety of valuable product dating tools available to them. Collectively, these are called day dot systems. Essentially, a day dot system consists of a variety of types and sizes of labels used to manage an operation's in-house product "Use-By" dates.
>
> To see examples of the various day dot systems available in the market and learn how they help control inventory costs by minimizing product loss, enter "day dot inventory management systems for restaurants" in a search engine and view the results.
>
> *www.fsis.usda.gov/food-safety/safe-food-handling-and-preparation/food-safety-basics/food-product-dating*

To maintain the quality and security of delivered food items, upon receiving they should be immediately placed into one of three storage areas:

1) Dry-storage
2) Refrigerated storage
3) Frozen storage

**Dry-Storage**

Dry-storage areas should be maintained at a temperature ranging between 65°F and 75°F (18°C and 24°C). Temperatures below that range can be harmful to food products. More often, however, dry-storage temperatures can increase and exceed by far the upper limit of temperature acceptability. This is because storage areas are frequently located in poorly ventilated and closed-in areas of an operation, and excessively high or low temperatures will damage dry-storage products.

Shelving in dry-storage areas must be easily cleaned and sturdy enough to hold the weight of dry products. Slotted shelving is preferred over solid shelving when storing food because slotted shelving allows for better air circulation around stored products.

All shelving should be placed at least six inches above the ground (floor) to allow for proper cleaning beneath the shelving and to ensure proper ventilation. Dry-goods products should never be stored directly on the floor. Dry-goods may not be stored:

✓ In locker rooms
✓ In toilet rooms
✓ In dressing rooms
✓ In garbage rooms
✓ In mechanical rooms
✓ Under sewer lines that are not shielded to intercept potential drips
✓ Under leaking water lines including leaking automatic fire sprinkler heads or under lines on which water has condensed
✓ Under open stairwells

When placed into storage, all the labels of cans and cases should face out for easy identification. Bulk items such as flour or sugar should be stored in wheeled bins when possible so that heavy lifting and the potential for employee injuries related to lifting can be avoided. Most importantly, dry-storage spaces must be sufficient in size to handle an operation's storage needs.

Cramped and cluttered dry-storage areas tend to increase costs because inventory cannot be easily rotated, maintained, located, or counted. Hallways leading to storage areas should always be kept clear and free of excess storage materials and empty boxes. This helps both in accessing the storage area and in reducing the number of potential hiding places for insects and other pests.

### Refrigerated Storage

Refrigerated storage is recommended for most **Time and temperature control for safety foods (TCS foods).**

In general, TCS foods are those which consist in whole or in part of milk or milk products, eggs, meat, poultry, rice, fish, shellfish, and edible crustaceans. They can also be raw-seed sprouts, heat-treated vegetables, and vegetable products and other ingredients in a form that supports rapid and progressive growth of microorganisms.

**Key Term**

**Time and temperature control for safety foods (TCS foods):** Foods that must be kept at a particular temperature to minimize the growth of food poisoning bacteria or to stop the formation of harmful toxins.

Refrigerator temperatures used to store TCS foods should be maintained at or below 40°F (4°C). In most cases, the lower areas tend to be coldest in refrigerators because warm air rises and cold air falls. In fact, the refrigerator itself may vary as much as several degrees (F) between its coldest spot (near the bottom) and its warmest spot (near the top). In most cases, it is advisable to set refrigerator thermostats at 39°F (3.9°C) to ensure proper cold-food holding temperatures are maintained.

Refrigerators should be opened and closed quickly when used, both to lower operational costs and to ensure that the items in the refrigerator stay at their peak quality. Refrigerators should be properly cleaned on a regular basis. Condensation drainage systems in refrigerators should be checked at least weekly to ensure they are kept clean and are functioning properly.

### Freezer Storage

Freezer temperatures should be maintained at 0°F (−18°C) or less. Newly delivered products should be carefully checked with a thermometer when received to ensure they are solidly frozen and have been delivered at the proper temperature. In addition, these items should be carefully inspected to ensure that they have not been thawed and then refrozen. This is because refrigerators and frozen-food holding units remove significant amounts of stored product moisture, causing shrinkage and **freezer burn** in meats and produce.

Unless they are built in, frozen-food holding units as well as refrigerators should be high enough off the ground (floor) to allow for easy cleaning around and under them and to prevent cockroaches and other insect pests from living beneath them. Stand-alone units should be placed six to ten inches away from walls to allow for the free circulation of air around, and efficient operation of, the units.

**Key Term**

**Freezer burn:** A form of deterioration in product quality resulting from poorly wrapped or excessively old items stored at freezing temperatures.

Frozen-food holding units must be regularly maintained, a process that includes cleaning them inside and out, and constant temperature monitoring to detect possible improper operation. A thermometer permanently placed in the unit, or one easily read from outside the unit, is best. It is also a good idea to periodically check that gaskets on freezers and refrigerators tightly seal the food cabinet. This will reduce operating costs while maintaining peak food quality for a longer time.

---

**Technology at Work**

Refrigerator and frozen-food holding units that operate within recommended temperature zones are essential for maintaining product quality and food safety. If proper temperatures are not maintained, the consequences can be serious. For example, if proper temperature control is not prioritized, *Salmonella, Campylobacter, Listeria,* and *E. coli* can contaminate cold foods.

Historically, foodservice operators have been advised to periodically take the temperatures of their cold-food holding units, or regularly monitor the thermometers that are built into the units themselves. That worked, of course, when staff were available to do the temperature monitoring, but it did not work when the operation was closed for hours or days, and on-site monitoring was not possible.

Today, foodservice operators can employ automatic temperature monitoring systems. These systems are designed to regularly monitor cold-food holding units, record the data obtained, and alert management to problems immediately with automatically generated text or e-mails to an operator's iPad, Android tablet, or other smart device. If cold-food holding unit temperatures fall below established standards, these inexpensive early warning systems alert operators to potential problems, and they can act before the loss of significant food products.

To learn more about automatic temperature monitoring systems available to foodservice operators, enter "restaurant cold storage monitoring systems" in your favorite search engine and view the results.

## Security Concerns

Inventoried items are the same as money to a foodservice operator, so they should be thought of in exactly that way. As previously noted, the New York strip steak in a walk-in is not just a steak. It represents, at a selling price of $20.00, that amount in revenue generated by the foodservice operator who hopes to sell a perfectly cooked steak to a hungry diner. If the steak disappears, revenue of $20.00 will disappear also. When operators think of inventory items in terms of their sales value, it becomes clear why product security is of the utmost importance.

All foodservice establishments will experience some amount of theft, at least in its strictest sense, and the reason is simple. Some employee theft is impossible to detect. Even the most sophisticated, computerized control system is not able to determine if an employee or supplier's employee entered the produce walk-in and ate one green grape. Similarly, an employee who takes home two small sugar packets per night will likely go undetected. In neither of these cases, however, is the amount of loss significant, and certainly not enough to install security cameras in the walk-in or to search employees as they leave for home at the end of their shifts.

What operators must do, however, is to make it difficult to remove significant amounts of food or beverage items from storage without authorization and to alert a manager when these items have been removed. Good cost control systems must be in place if this goal is to be achieved, and profits are to be optimized.

To illustrate the impact of product theft on profitability, consider Jesse, an employee of the Irish Times food truck. On a daily basis, Jesse takes home $3.00 worth of food products. How much, then, does Jesse cost this food truck's owner in one year? The answer is a surprising $21,900 in sales revenue!

If Jesse or others pilfer $3.00 per day for 365 days, the total theft amount would be $1,095 (365 days × $3.00 per day = $1,095). If the food truck makes an after-tax profit of 5 percent on each dollar of food products sold that is needed to recover the lost $1,095, the operation must make additional annual sales of ($1.00/0.05) × $1,095 = $21,900! In the case of a smaller operation, $21,900 may well represent several days' or even weeks of sales revenue. Clearly, small thefts can add up to large dollar losses!

When developing an inventory security program, foodservice operators must attempt to carefully control access to the locations of their stored products. In some operations, this may be done by a process as simple as keeping the dry-storage area locked. Then, employees must "get the keys" from a manager or supervisor when products are needed.

In other operations, cameras may be mounted in both storage areas and employee exit areas. Sometimes, the physical layout of a foodservice operation may prevent management from being able to effectively lock and secure all storage areas, but too easy access is frequently a cause of theft problems. This is not because employees are basically dishonest, and most are not! Theft problems develop because of the few employees who feel either that management will not miss a minimal quantity of whatever is being stolen, or they falsely feel they "deserve" to take a few things because they work so hard.

It is management's responsibility to ensure that product inventories remain secure. Generally, if storerooms are to be locked, only one individual should have access to the storage areas during any shift. In reality, however, it may be difficult to keep all inventory items under lock and key. For example, some items must be received and immediately sent to the kitchen for processing or use. As a result, most operators find that it is impossible to operate under a system where all food products are locked away from all employees. Storage areas should, however, not be accessible to guests or vendor employees.

If proper control procedures are in place, employees will know that management can determine if theft has occurred. Without such control, employees may feel that theft will go undetected, and this must be avoided.

## Valuing Product Inventories

It is important for operators to continually monitor the amount of products they have in inventory. Inventory levels may be determined by counting stored items, as in the case of cans, or by weighing items, as in the case of meats. Volume (e.g., gallons and quarts) is another method of establishing product amounts. Other purchase units (see Chapter 4) such as bushels, crates, flats, or bags may also be used to establish product amounts.

If an item is purchased by the pound, it is generally weighed to determine the amount on hand. If it is purchased by the piece or case, the appropriate unit to determine the item amount may be either pieces or cases. If, for example, canned pears are purchased by the case, with six #10 cans per case, an operator might decide to count the item in terms of either cases or cans. That is, three cases of the canned pears might be considered as three items (when counted by the case) or as 18 items (when counted by the can). Properly used, either method can correctly establish the amount of product an operation has on hand and is acceptable.

Foodservice operators must know the amount of product inventory they have on hand, but they must also know the monetary value of their stored items. An operation's current inventory values are needed to prepare its **balance sheet**, where inventories are listed as a current asset.

Inventory values are also necessary for the proper preparation of an operation's income statement (see Chapter 1) and inventory values are needed to establish an accurate "cost of goods sold" amount.

Proper inventory control requires operators to monitor both the amount and value of items in product inventories. Valuing, or establishing a dollar value for an inventory item, is performed using the following item inventory value formula:

**Key Term**

**Balance sheet:** A report that documents the assets, liabilities, and net worth (owners' equity) of a foodservice business at a single point in time. Also commonly called the Statement of Financial Position.

$$\text{Item Amount} \times \text{Item Value} = \text{Item Inventory Value}$$

An item's actual inventory value can often be more complicated to determine than its amount (quantity). This is because the price an operator pays for an item may vary slightly each time it is purchased.

For example, assume that an operator bought curly endive for $4.20 per pound on Monday, but the same item costs $5.50 per pound when it is delivered on Wednesday. On Friday, the operator sees that they have two pounds of the curly endive in their refrigerated walk-in. Is the value of this item $8.40 (2 pounds × $4.20 per pound = $8.40) or $11.00 (2 pounds × $5.50 per pound = $11.00)?

Item value can be determined by using either the LIFO or the FIFO methods. When the LIFO method is in use, the item's value is said to be the price paid for the least recent (oldest) addition to inventory.

If the FIFO method is in use, the item value is said to be the price paid for the most recent (newest) product on hand. In the hospitality industry, most operators value inventory at its most recently known (current) value. Therefore, FIFO is the more commonly used inventory valuation method.

Inventory value is determined using a form or spreadsheet similar to the inventory valuation sheet shown in Figure 5.2. This form has a place for entering all inventory items, the quantity of each item on hand, and the item (purchase unit) value of each. There is also a place for the date the inventory was taken, a position for the name of the person who counted the product, and another area for the person who extends (calculates) the monetary value of the inventory. It is

Operation Name: _____     Inventory Date: _____

Counted By: _____     Extended By: _____

| Item | Unit | Item Amount | Item Value | Inventory Value |
|------|------|-------------|------------|-----------------|
|  |  |  |  |  |
|  |  |  |  |  |
|  |  |  |  |  |
|  |  |  |  |  |
|  |  |  |  |  |
|  |  |  |  |  |
|  |  |  |  |  |
|  |  |  |  |  |
|  |  |  |  |  |
|  |  |  |  |  |
|  |  |  |  |  |
|  |  |  |  |  |
|  |  |  |  |  |
|  |  |  |  |  |
|  |  |  |  |  |
|  |  |  |  |  |
|  |  |  |  |  |
|  |  |  |  |  |
|  |  |  |  |  |
|  |  |  |  |  |
|  |  |  |  |  |
|  |  |  |  |  |
|  |  |  | Page Total |  |

Page _____ of _____

**Figure 5.2**   Inventory Valuation Sheet

recommended that, when possible, these be two different individuals to reduce the risk of inventory valuation fraud.

An inventory valuation sheet should be completed each time the inventory is taken (counted). It can be manually prepared or produced as part of an inventory evaluation software program.

Regardless of the system used, each item's total inventory value is determined using the inventory value formula. For example, if an operation has five cases of fresh beets in inventory, and each case has a value of $25.00, the inventory value of beets would be:

Item Amount × Item Value = Item Inventory Value

or

5 cases × $25.00 per case = $125

---

**Technology at Work**

Those foodservice operators selling alcoholic beverages know that determining the inventory value of opened and partially used liquor bottles can be challenging and is often quite time consuming. Variations in bottle sizes and shapes can make it difficult to identify how much product is actually in an open bottle and, of course, knowing the amount in the bottle is essential for determining its inventory value.

In the past, various methods have been proposed to address this issue including weighing bottles and then subtracting the bottles' weight or estimating content amounts based on a tenth system. In such a system, bottles are visually inspected and then their contents are estimated in a series of tenths (i.e., 1/10th, 5/10th, 8/10th, and the like). In this system, operators then multiply the value of a full bottle by the appropriate number of tenths it contains to determine the inventory value of the partially used bottle.

Today's foodservice operators have better choices. A variety of mobile apps have been developed specifically to help operators manage and control their bar inventories. These programs can be downloaded to a smartphone or tablet to provide easy access to an automated bar inventory management system. The capabilities of the apps vary by manufacturer. However, in most cases, the systems allow operators to view a display of virtual bottles, then tap the screen at the alcohol level in each bottle or use a smart camera to take a picture of the bottle and allow for automatic product value determination. The best of such systems is interfaced with an operation's point-of-sale (POS) system, which then also allows for a comparison between beverages sold and beverage inventory used. This, in turn, results in the development of detailed inventory variance reports for each individual liquor product.

To learn more about these useful inventory management programs, enter "bar inventory apps" in your favorite search engine and review the results.

## Issuing and Restocking Product Inventories

The process of determining inventory values requires that a member of an operation's staff count all inventoried products on hand and multiply the value of the item by the number of units on hand. The process becomes more difficult when one realizes that the average foodservice operation has hundreds of items in inventory. Thus, "taking" the inventory can be a very time-consuming task. A **physical inventory** must, however, be taken to determine an operation's actual product usage and to help identify the need for purchasing (restocking) inventory items.

Some operators take a physical inventory monthly, others weekly or even daily, to determine their inventory amounts on hand. Frequency of inventory as an aid to calculating actual food and beverage usage and their product cost percentages (see Chapter 6), and to determine whether additional products are needed, varies by property.

**Key Term**

**Physical inventory:** An inventory control system tool in which an actual (physical) count and valuation of all product inventory on hand is taken at the close of an accounting period.

---

**What Would You Do? 5.1**

"I know what Randy asked me to do, I'm just not sure how to do it!" said Sara.

Sara, the assistant manager at the Sombrero Mexican Restaurant, was talking to Ollie, the restaurant's dining room manager.

"What's the problem?" asked Ollie.

"The problem," replied Sara, "is that Randy asked me to do the weekly inventory while he's on vacation. I'm supposed to count the items we have in inventory, multiply the amount we have on hand by the cost of the items, and determine our total inventory value."

"That seems easy enough," said Ollie.

"Well the counting part is easy," said Sara, "and I certainly know how to multiply. What I don't know is how much to multiply by. For example, we have 80 pounds of the skirt steaks we use for Fajitas in inventory. Some of it was purchased at $16 a pound and some at $20 a pound."

"So maybe multiply by $18 a pound?" questioned Ollie.

**Assume you were the owner of the Sombrero Mexican Restaurant. How important do you think it would be for inventory valuations produced by Sara to be calculated in the same way as those previously taken by Randy? What would be the likely result if they were not?**

## Issuing

**Issuing** is the process by which needed food and beverage products are removed from product inventory. In small foodservice operations, issuing may be as simple as entering a locked storeroom, selecting products, and relocking the door upon leaving. In a more complex operation, especially one that serves alcoholic beverages, this method may be inadequate to achieve appropriate product control.

In the ideal case, the issuing of products should occur only after the product has been requisitioned. A **requisition** is simply a formal request to have products issued from storage.

The act of requisitioning products from the storage area is important, but it does not need to be unduly complex. Sometimes, foodservice operators create difficulties for their workers by developing a requisition system that is far too time-consuming and complicated. The difficulty in such an approach usually arises because management hopes to equate products issued from storage with products sold without taking a physical inventory. In reality, this process is difficult, if not impossible, to carry out.

To better understand why, consider the operator whose business includes a popular bar area. If, on any given night, the operator attempted to match liquor issued to liquor sold, this operator would need to assume that all liquor issued on a given day is sold on the same day and that no liquor issued on a prior day was sold on the given day. This, of course, will not be the case. Similarly, in the kitchen, some items issued today (e.g., ice cream) may be sold over several days.

It is good management to view an issuing system as one of providing basic product security and to view a total inventory control/cost system as a separate entity entirely.

Employees should always requisition food and beverage items based on management-approved estimates of future sales and product production schedules. Although special care must be taken to ensure that employees use the products for their intended purpose, maintaining product inventory security can be achieved with relative ease if a few key principles are observed:

1) Food, beverages, and supplies should be requisitioned only as needed based on approved production schedules.
2) Needed items should be issued only with management approval.

**Key Term**

**Issuing:** The process of transferring food and beverage products from storage to production areas for use in a foodservice operation.

**Key Term**

**Requisition:** The formal act of requesting a food or beverage product to be removed from storage for use in an operation.

3) If a written record of issues is to be kept, each person removing food, beverages, or supplies from the storage area must sign to acknowledge receipt of the products.
4) Products that are not ultimately used should be returned to their proper storage area, and their return should be recorded to increase the inventory level as appropriate.

Regardless of the methods used by employees to requisition food and beverage products or the systems management uses to issue these products, product inventory levels in an operation will be reduced as sales are made.

It is the responsibility of the individual in charge of purchasing to monitor reductions in product inventories and purchase additional products as needed. The proper restocking of inventory is critical if product shortages are to be avoided and if products needed for menu item preparation are to be available.

Nothing is quite as traumatic for a foodservice operator as being in the middle of a busy meal period and finding that the operation is "out" of a necessary ingredient or frequently requested menu item. For most operators, it would be very time-consuming to monitor the amount of each ingredient or product in inventory on a daily basis. The average foodservice operation stocks hundreds of ingredients and items, each of which may or may not be used every day. The task could be overwhelming.

Consider, for example, the difficulty associated with daily monitoring the use of each catsup packet provided to guests of a high-volume quick-service restaurant. Taking a daily physical inventory for the use of such a product would be similar to spending $10 to protect a penny! The effective foodservice operator knows that proper control involves spending time and effort where it is most needed and where it will do the most good. It is for this reason that many operators practice the ABC method of inventory control.

To best understand the principles utilized by the ABC method of inventory product control, operators must first understand the advantages and disadvantages of the previously addressed physical inventory and a **perpetual inventory system**.

A physical inventory system is one in which an actual physical count and valuation of all inventories on hand is taken at the close of an accounting period. In contrast, a perpetual inventory is one in which the entire inventory is counted and recorded, and then additions to and deletions from total inventory are recorded as they occur.

**Key Term**

**Perpetual inventory system:** An inventory control system in which additions to and deletions from total inventory are recorded as they occur.

Both physical and perpetual inventory systems have advantages and disadvantages for the foodservice operator. The physical inventory, properly taken, is the most accurate of all because each item is actually counted and then valued. It is the physical inventory, taken at the end of the accounting period (the ending inventory), that is used in conjunction with the beginning inventory (the ending inventory value from the prior accounting period) to compute an operation's cost of food or beverages sold (see Chapter 6).

Despite its accuracy, however, the physical inventory suffers from a significant disadvantage: It can be extremely time-consuming to complete. Even when software programs that extend inventory (multiply the number of purchase units on hand by each unit's value) or handheld code scanners are used, counting each item in storage can be time-consuming. This is so even for the trained individuals who carefully weigh/count the inventoried items.

A perpetual inventory system seeks to eliminate the need for frequent counting by adding to the inventory when appropriate (via delivery invoices) and subtracting from inventory when appropriate (via recorded requisitions or issues). Perpetual inventory systems are especially popular in monitoring liquor and wine inventories when each product has its own inventory sheet or, in some operations, a **bin card**.

When using electronic or hard copy bin cards to monitor inventory levels, operators must only review the cards, rather than take a physical inventory, to determine the quantity of products that should be available at any point in time.

### Key Term

**Bin card:** A line on a spreadsheet (or a physical card) that details additions to and deletions from a product's inventory level.

However, this will be true only when every addition to and every subtraction from products in inventory are carefully recorded.

Today, foodservice operators increasingly use computer spreadsheets and specialized inventory software to maintain perpetual inventories. In addition, many modern POS systems include built-in inventory management components.

The accurate use of a perpetual inventory system requires that each change in product quantity be noted. However, employees, when in a hurry, may simply forget to update the perpetual system as they add or remove inventory items. Mistakes such as these will reduce the accuracy of the perpetual inventory. For this reason, it is not wise to depend solely on a perpetual inventory system for accurate cost calculations. There are, however, several advantages to the perpetual inventory system, among them the ability of the purchasing agent to quickly note the quantity of product on hand without resorting to a daily physical inventory count.

The question about which of the two inventory systems is best arises when deciding whether to use a physical or perpetual inventory system. Experienced operators know that neither is best in all cases, so they select the best features of both systems as needed by their own operations. For example, they may use a perpetual system for "expensive" products and a physical inventory system for their less-expensive counterparts.

## ABC Inventory Control

Foodservice operators seeking to utilize the best of both the physical and perpetual inventory systems may elect to implement an **ABC inventory control system.**

Utilizing the most useful features of physical and perpetual inventory systems is what the ABC inventory system is designed to do. It requires operators to separate their inventory items into three main categories:

**Key Term**

**ABC inventory control system:** An inventory control system utilizing aspects of both physical and perpetual inventory systems.

*Category A* items are those that require tight control and the most accurate record-keeping. These are typically high-value items and, though few in number, they can make up 70–80% of an operation's total inventory value.

*Category B* items are those that make up 10–15% of the inventory value and require only routine control and recordkeeping.

*Category C* items make up only 5–10% of the inventory value. These items require only the simplest of inventory control systems.

For a simple example of the use of the ABC system, assume that the following 10 items are routinely held in an operation's product inventory to prepare a menu item:

1) Precut New York strip steaks
2) Prepared horseradish
3) Eight-ounce chicken breasts (fresh)
4) Garlic salt
5) Onion rings
6) Crushed red pepper
7) Dried parsley
8) Lime juice
9) Fresh tomatoes
10) Rosemary sprigs

As can be seen even with this short list, the operation has a variety of items in inventory. Some of these items, like the New York strip steak, are very valuable, highly perishable, and critical for the execution of the operation's menu.

Other items, like the crushed red pepper, are much less costly, not highly perishable, and may less dramatically affect the operation if it ran out between deliveries.

When using the ABC system, items such as New York strip steaks and crushed red pepper are not treated the same because they are not equally critical to the operation's success. The ABC system helps operators distinguish between those items that deserve special, perhaps daily, attention, and other items that do not require careful scrutiny.

To develop the A, B, and C inventory categories, operators:

1) Calculate monthly usage in units (e.g., pounds, gallons, and cases) for each inventory item.
2) Multiply total item usage times its purchase unit price (unit value) to arrive at the total monthly amount of product usage.
3) Rank items from highest dollar usage to lowest.

In a typical ABC analysis, approximately 20% of the items held in inventory will account for about 70% to 80% of the total monthly product cost. These represent the A product category. It is not necessary that the line separating A, B, and C products be drawn the same for every operation, but many operators use the following simple guide:

Category A—Top 20% of items
Category B—Next 30% of items
Category C—Next 50% of items

It is important to note that, although the *percentage* of items in category A is small, the percentage of total monthly product costs these items represent is large. Alternatively, the number of items in category C is large, but the total dollar value of product cost these items account for is rather small. The ABC inventory system is concerned with the monetary value of products rather than the number of items.

Figure 5.3 shows the complete result of performing an ABC analysis on the 10 ingredients listed previously and then ranking those items in terms of their inventory value.

| Item | Monthly Usage | Purchase Price ($) | Monthly Usage ($) | Category |
|---|---|---|---|---|
| Precut New York strip steaks | 300 lb. | 12.50/lb. | 3,750.00 | A |
| 8-ounce chicken breasts (fresh) | 450 lb. | 4.10/lb. | 1,845.00 | A |
| Fresh tomatoes | 115 lb. | 0.95/lb. | 109.25 | B |
| Onion rings | 30 lb. | 2.20/lb. | 66.00 | B |
| Rosemary sprigs | 10 lb. | 4.50/lb. | 45.00 | B |
| Prepared horseradish | 4 lb. | 2.85/lb. | 11.40 | C |
| Lime juice | 2 qt. | 4.10/qt. | 8.20 | C |
| Garlic salt | 2 lb. | 2.95/lb. | 5.90 | C |
| Crushed red pepper | 1 oz. | 16.00/lb. | 1.00 | C |
| Dried parsley | 4 oz. | 4.00/lb. | 1.00 | C |

**Figure 5.3** ABC Inventory Analysis on Selected Inventory Items

| Category | Inventory Management Techniques |
|---|---|
| A | 1) Order only on an as-needed basis. |
| | 2) Conduct perpetual inventory on a daily or at least weekly basis. |
| | 3) Have clear idea of the purchase point and estimated delivery times. |
| | 4) Conduct monthly physical inventory. |
| B | 1) Maintain normal control systems; order predetermined inventory (par) levels. |
| | 2) Monitor more closely if the use of this item is tied to sale of an item in category A. |
| | 3) Review status quarterly for movement to category A or C. |
| | 4) Conduct monthly physical inventory. |
| C | 1) Order in large quantity to take advantage of discounts if item is not perishable. |
| | 2) Stock consistent levels of product. |
| | 3) Conduct monthly physical inventory. |

**Figure 5.4** Guide to Managing ABC Inventory Items

The ABC inventory system specifically directs an operator's attention to the areas where it is most needed. Controlling product costs, especially for category A items, is extremely important. Figure 5.4 details the differences in how an operator might manage items in the A, B, and C inventory categories.

The ABC system focuses management's attention on the essential few items in inventory, and it focuses less attention on the many low-cost, slower-moving items. Again, it is important to note that management's time is best spent on the items of most importance.

The ABC system can also be used to arrange storerooms or to determine which items should be stored in the most secure areas. Regardless of the inventory management system used, however, whether it is the physical, perpetual, or ABC inventory, management must be strict in monitoring both withdrawals from inventory and the process by which inventory is restocked.

### Restocking Product Inventories

When product inventories are properly controlled, and issues from inventories are carefully recorded, foodservice operators will be in an excellent position to determine when product inventories should be restocked. They can do so by creating a **purchase point** for each item held in inventory.

**Key Term**

**Purchase point:** The inventory level at which a stored item should be reordered.

In most cases, an item's purchase point can be identified by one of two methods:

1) As needed
2) Par level

### As Needed

When operators select the **as-needed** (just-in-time) method of determining an item's purchase point, they are basically purchasing food based on their prediction of future sales (see Chapter 2) and the amount of the ingredients (from standardized recipes) necessary to produce those sales. Then, no more than the absolute minimum of needed inventory level (plus safety stock: see Chapter 3) is purchased from suppliers.

When the as-needed system is used, an operator compiles a list of needed ingredients or products to purchase. Then, that amount, and no more, would be ordered from the appropriate suppliers.

**Key Term**

**As needed (purchase point):** A system of determining the purchase point by using sales forecasts and standardized recipes to determine how much of an item should be held in inventory. Also referred to as "just-in-time."

### Par Level

Foodservice operators may also set predetermined purchase points, called **par levels**, for some inventory items. Par levels are best used for items that have a relatively long shelf life (see Chapter 3) and that can be purchased in bulk.

**Key Term**

**Par level:** The amount of a stored item that should be held in inventory at all times.

For example, all foodservice operators have standardized recipes that call for salt as an ingredient. It does not make sense, however, to order salt by the number of teaspoons or tablespoons that are required. In instances such as this, or when demand for a product is relatively constant, an operator may decide that the use of an "as-needed" ordering system will not work as well as would identifying appropriate par levels for some products or ingredients.

When determining appropriate par levels, operators establish minimum and maximum amounts to be held in inventory. Many foodservice managers establish a minimum par level by computing working stock and then adding 25 to 50 percent more for safety stock. Then, an appropriate purchase point, or point at which additional stock is purchased, is determined.

If, for example, an operator decides that the inventory level for coffee should be based on a par-level system, the decision may be made that the minimum amount (based on past usage) of coffee that should be on hand at all times is four cases. This would be the minimum par level.

To ensure the operation never ran out of coffee, however, assume this operator set the maximum par level as ten cases. Although the actual inventory level in this scenario would vary from a low of four cases to a high of ten cases, this operator would be assured that there would never be too little or too much coffee in inventory.

When cases of coffee are ordered under this system, the operator always attempts to keep the number of cases on hand between the minimum par level (four cases) and the maximum par level (ten cases). The purchase point in this example might be six cases; that is, when the operation had six cases of coffee on hand, an order would be placed with the coffee vendor. The intention would be to get the total stock of coffee up to ten cases before the supply in inventory got below four cases. Delivery might take one or two days so, depending on supplier delivery schedules, six cases might be an appropriate purchase point.

Whether an operation uses an as-needed, a par-level, or, as in the case of many operators, a combination of both systems, each ingredient or menu item should have a pre-set and management-designated inventory level or amount.

As a rule, most highly perishable items should be ordered on an as-needed basis, and items with a longer shelf life can often have their inventory levels set using a par-level system. The answer to the question "How much of each ingredient should an operation have on hand at any point in time?" must come from an operator's estimate of future sales.

---

**Find Out More**

Inventory management software can help foodservice operators manage their inventory levels, purchase orders (POs), standardized recipes, and item costs. The best of such systems are integrated with an operation's POS system and can save hours of administrative work.

Essentially, in these programs, operators enter their standardized recipes into the system, and the POS system creates product usage summaries based on sales as recorded in the POS system. The systems can then generate estimated POs based on an operation's actual item sales. They can also produce inventory usage summaries.

To see some examples of these helpful programs, enter "restaurant inventory software systems" in your favorite search engine and view the results.

---

The proper purchasing, receiving, storage, issuing, and restocking of foodservice products is essential if an operation is to control its product costs. When all required activities have been properly performed, a foodservice operation's production staff will be well prepared to produce the items an operation sells to its guests. Cost control during this critical production process is essential, and for that reason controlling product costs during production is the topic of the next chapter.

**What Would You Do? 5.2**

"This doesn't make any sense," said Virginia, "how can we be short five steaks in only two days?"

"What do you mean we're five steaks short?" replied Vernon, "short of what?"

Virginia was the manager and food buyer at Lucky's Steakhouse. She was talking to Vernon, one of the restaurant's daytime prep cooks. It was 8:00 o'clock on Monday morning, and Virginia had just inspected the walk-in coolers where the four different types of steaks featured by the restaurant were stored.

"Short of what we should have," said Virginia. "When I left on Saturday morning, we had 44 of our 16 oz. rib eyes in the walk-in. I counted them and wrote it down. According to the POS we sold 18 of them on Saturday night and six more on Sunday night, for a total of 24."

"So," said Vernon, "what's wrong with that?"

"Well, if we started with 44 and sold 24, we should still have 20 steaks on hand. But I have counted them twice this morning and we only have 15!"

"Maybe the POS is wrong?" offered Vernon.

**Assume you were Virginia. What are some reasons why the number of steaks remaining in inventory would be less than those that would be indicated by the operation's POS system? What steps could you take to identify and correct the issues that could have caused this shortage?**

## Key Terms

Product inventory

LIFO (last-in first-out) inventory system

FIFO (first-in first-out) inventory system

QR code

Time and temperature control for safety food (TCS food)

Freezer burn

Balance sheet

Physical inventory

Issuing

Requisition

Perpetual inventory system

Bin card

ABC inventory control system

Purchase point

As needed (purchase point)

Par level

## Operator's 10-Point Tactics for Success Checklist

Evaluate your need for, and the current status of, each of the following operational tactics. For those tactics you think are important, but not yet in place, develop an action plan for its implementation including who will be responsible for the tactic's completion and the target date by which it should be completed.

| Tactic | Don't Agree (Not Done) | Agree (Done) | Agree (Not Done) | If Not Done | |
|---|---|---|---|---|---|
| | | | | Who Is Responsible? | Target Completion Date |
| 1) Operator recognizes the importance of proper storage and issuing practices in the control of product costs. | ___ | ___ | ___ | | |
| 2) Operator can list key principles for protecting the quality of food items held in dry-storage areas. | ___ | ___ | ___ | | |
| 3) Operator can list key principles for protecting the quality of food items held in refrigerators and freezers. | ___ | ___ | ___ | | |
| 4) Operator recognizes the importance of and need for security measures taken to protect product inventories from theft. | ___ | ___ | ___ | | |
| 5) Operator can state the reasons a physical inventory must be taken to accurately establish the monetary value of products in storage. | ___ | ___ | ___ | | |
| 6) Operator can state the advantages of using a perpetual inventory system for selected stored products. | ___ | ___ | ___ | | |
| 7) Operator can state the advantages of using an ABC inventory system to monitor product inventory levels. | ___ | ___ | ___ | | |

| Tactic | Don't Agree (Not Done) | Agree (Done) | Agree (Not Done) | If Not Done | |
|---|---|---|---|---|---|
| | | | | Who Is Responsible? | Target Completion Date |
| 8) Operator can conduct the analysis needed to classify stored items as either A, B, or C when using the ABC inventory system. | _____ | _____ | _____ | | |
| 9) Operator understands how to restock inventory items using an "as-needed" approach to item purchasing. | _____ | _____ | _____ | | |
| 10) Operator understands how to restock inventory items using a "par-level" approach to item purchasing. | _____ | _____ | _____ | | |

# 6

# Cost Control in Food and Beverage Production

---

**What You Will Learn**

1) The Importance of Cost Control in Food Production
2) The Importance of Cost Control in Beverage Production
3) How to Manage Standardized Recipes
4) How to Calculate Actual Cost of Sales

---

**Operator's Brief**

In this chapter, you will learn that managing product costs during production will be extremely important to the success of your foodservice operation. You will learn specific ways to manage your product costs during food and beverage production.

One extremely important aspect of cost control during production relates to the use and management of standardized recipes. A standardized recipe should be prepared for each menu item sold. In many cases, however, you will need to adjust a standardized recipe to produce the number of servings you wish to prepare. In this chapter, you will learn how to adjust (scale) standardized recipe yields using a recipe conversion factor (RCF).

The proper management of standardized recipes also includes the ability to accurately "cost" a recipe to determine the recipe's total food cost and its per portion cost. Doing so requires an understanding of "as" purchased ("AP") costs and edible portion ("EP") costs. In this chapter, you will learn how to perform yield tests to calculate waste percentages and yield percentages of recipe ingredients. With this task completed, the actual costs of these ingredients can be utilized to calculate accurate, standardized recipe portion costs.

Finally, at the conclusion of the chapter, you will learn how to calculate your operation's actual cost of sales for food and beverage, and how to calculate actual cost of sales for food and beverages when products are transferred to, or from, a kitchen or bar. These amounts must be accurately calculated because they are important components of the income statements you will prepare at the conclusion of each of your financial reporting periods.

---

**CHAPTER OUTLINE**

The Importance of Cost Control During Production
    Managing Costs in Food Production
    Managing Costs in Beverage Production
Controlling Cost Through Standardized Recipes
    Adjusting Standardized Recipe Yields
    Costing Standardized Recipes
Calculating Actual Cost of Sales
    Calculating Actual Cost of Food Sold
    Calculating Actual Cost of Beverage Sold
    Calculating Actual Cost of Sales with Transfers

---

# The Importance of Cost Control During Production

When professional procedures have been implemented to properly purchase, receive, and store products, foodservice operators will be in a good position to continue their cost control efforts during the production process.

For those operators who sell both food and alcoholic beverage products, cost control efforts will focus on:

✓ Managing costs in food production
✓ Managing costs in beverage production

## Managing Costs in Food Production

Effective management of the production process is essential for effective cost control. In most cases, managing the production process in a foodservice operation means controlling four key areas:

✓ Waste
✓ Overcooking
✓ Over-portioning
✓ Improper carryover utilization

### Waste

Product losses from food (and beverage) waste is one example of excessive food costs. Some waste may be easy to observe such as when an employee does not use a rubber spatula to remove all salad dressing from a 1-gallon jar. However, acts such as the improper work of a salad preparation person can yield excessive amounts of trim waste and, as a result, waste a higher **portion cost** for each salad sold.

Portion cost is the amount it costs an operation to produce one serving of a menu item. Portion cost is one important factor in establishing the selling prices of menu items, but it is also important for the control of production costs. For that reason, the ability to calculate a portion cost correctly is an important management skill.

In some cases, the calculation of a portion cost is very easy. For example, assume an operator sells fresh apples. The operator buys apples in 10-pound bags that contain 30 apples. Each bag costs $18.90. The operator could easily calculate the portion cost for one apple:

$$\frac{\$18.90 \text{ apple cost per bag}}{30 \text{ apple portions per bag}} = \$0.63 \text{ portion cost}$$

In other situations, operators utilize the cost required to produce a standardized recipe when they calculate their portion costs.

For example, assume an operator sells pecan pie. The operator's standardized recipe for pecan pie produces seven pies. Each pie will be cut into eight pieces. Thus, the standardized recipe produces 7 pies × 8 portions per pie = 56 pie portions.

If the cost of producing the standardized recipe, including the pastry for the crust, pecans, eggs, brown sugar, salt, vanilla, milk, and nutmeg, is $50.40, the operator could again use the same portion cost formula to calculate the cost of producing one slice of pecan pie:

$$\frac{\$50.40 \text{ standardized recipe cost}}{56 \text{ pie portions produced}} = \$0.90 \text{ portion cost}$$

Since most items served by restaurateurs should always be produced by following standardized recipes, it is easy to see why adherence to proper recipe production and portioning is critical to accurately calculating and controlling an operation's portion costs.

Foodservice operators can best control food production costs by consistently demonstrating their concern for the value of food products. Each employee should realize that wasting products directly affects the operation's profitability and, therefore, the employees' own economic well-being.

In most cases, food waste results from poor training or an operator's inattentiveness to detail. Unfortunately, some operators and employees feel that small amounts of food waste are unimportant. Effective operators know that a primary goal in reducing production waste should be to maximize product utilization and minimize the dangerous "it's only a few pennies, and it doesn't matter" syndrome.

### Overcooking

Cooking is the process of exposing food to heat. Excessive cooking, however, most often results in reduced product volume regardless of whether the item cooked is roast beef or vegetable soup. The reason: Many foods have high moisture contents, and heating usually results in moisture loss. To minimize this loss, cooking times and methods included in standardized recipes must be calculated and followed.

Excess heat is the enemy of both well-prepared foods and an operator's cost control efforts. Too much time on the steam table line or in the holding oven extracts moisture from products, reduces product quality, and yields fewer portions to be available for sale. The result is unnecessarily increased portion costs and total food costs.

To control product loss from overcooking, operators must strictly enforce standardized recipe cooking times. This is especially true for high moisture content items such as meats, soups, stews, and baked goods. Moreover, extended cooking times can result in total product loss if properly prepared items are placed in an oven, fryer, steam equipment, or broiler and are then "forgotten" until it is too late! It is, therefore, advisable to provide kitchen production personnel with small, easily cleanable timers and thermometers for which they are responsible. These can help substantially in reducing product losses from overcooking.

Figure 6.1 illustrates the impact on portion costs of overcooking roast beef, a product used by many foodservice operations. In this example, a properly cooked roast will yield 100 8-ounce portions at a portion cost of $4.00. The portion cost of improperly (overcooked) roast beef ranges from $4.12 to $4.71!

It is also important to note that improperly cooked foods typically result in lower product quality, which results in decreased guest satisfaction.

### Over-portioning

Perhaps no other area of food and beverage cost control has been analyzed and described as fully as control of portion size, and there are two reasons for this. First, over-portioning by service personnel increases operating costs and may lead to mismatches in production schedules and anticipated demand. For example, assume 100 guests are expected, and 100 products to be served to them are

**Portion Cost of 50 lbs. (800 Ounces) of Roast Beef**

| Preparation State | Ending Weight (oz.) | 50 lb. Cost = $400<br>Number of 8-Ounce Portions | Portion Cost |
|---|---|---|---|
| Properly prepared | 800 | 100 | $4.00 |
| Overcooked 15 min. | 775 | 97 | 4.12 |
| Overcooked 30 min. | 750 | 94 | 4.26 |
| Overcooked 45 min. | 735 | 92 | 4.35 |
| Overcooked 60 min. | 720 | 90 | 4.44 |
| Overcooked 75 min. | 700 | 88 | 4.55 |
| Overcooked 90 min. | 680 | 85 | 4.71 |

**Figure 6.1**   Effect of Overcooking on Portion Cost

produced. However, over-portioning causes the operation to be "out" of the product after only 80 guests were served. The remaining 20 guests will not receive portions because over-portioning allowed their food to be served to other guests.

Secondly, over-portioning must be avoided because guests want to believe they receive fair value for what they spend. If portions are large one day and small the next, guests may believe they have been cheated on the second day. Consistency is a key to operational success in foodservice, and guests want to know exactly what they will receive for the money they spend.

In many cases, inexpensive kitchen tools are available to help employees serve the proper portion size. Whether these tools consist of scales, food scoops, ladles, dishes, or spoons, employees require an adequate number of easily accessible portion control devices and use them consistently. The constant checking of portion sizes served is also an essential management task. When incorrect portion sizes are noticed, they must be promptly corrected to avoid product cost increases.

### Improper Carryover Utilization
Most foodservice operators want to offer the same broad menu to the day's last guest as was offered to its first guest, so it is inevitable that some prepared food will remain unsold at the end of the shift (this food is referred to as "carryovers," or "leftovers").

In some segments of the hospitality industry, carryovers are a potential problem area, but less so in other operations. Consider the operator of an upscale gelato shop. At the end of the day, any unsold gelato is simply held in the freezer until the next day with no measurable loss of either product quantity or quality.

Contrast that situation, however, with a full-service cafeteria. If closing time is 8:00 p.m., management wishes to have a full product line, or at least some of each menu item, available to the guest who walks in the door at 7:55 p.m. Obviously, in five more minutes, many displayed items will become carryovers, and this cannot be avoided. An operator's ability to effectively integrate carryover items on subsequent days can make the difference between profits and losses in some operations.

In many cases, carryover foods cannot be sold for their original value. For example, today's beef stew made from yesterday's prime rib will not likely be sold at prime rib prices. Carryovers generally mean reduced income and less profits relative to product value, and it is critical to minimize carryovers.

Carryover items that can be re-used should be properly labeled, wrapped, and FIFO-rotated so the items can be found and re-used easily. This results in greater employee efficiency to locate and utilize carryover items along with reduced energy costs because refrigerator doors will be left open for shorter time periods. Most operators find that requiring foods to be properly labeled and stored in clear plastic containers helps to manage these procedures.

---

### What Would You Do? 6.1

"How many dozen should I put in the proofer?" asked Ezra, the new baker at the Old Country Cafeteria.

Fahad was the day shift operations manager and, unfortunately, he did not know how to answer Ezra's question. What Ezra wanted to know was simple enough: "How many dozen rolls should be placed in the proofer to anticipate the night's dinner business?"

The problem was that the frozen dinner roll dough used at the Old Country Cafeteria needed to proof for at least two hours before baking for 15 minutes.

If too many rolls were proofed, they would not be needed, but they would still have to be baked and made into bread dressing or even tossed away.

If too few dozen rolls were proofed and the night was busier than anticipated, they would run out of "Fresh Baked Rolls" (one of the operation's signature items), and Fahad knew that the night manager would be really upset. It was a daily guess, and sometimes Fahad missed the guess!

**Assume you were the owner of the Old Country Cafeteria. How important to your cost control efforts do you think it would be for Fahad to have accurate information about the night's sales forecast? Would you prefer that Fahad over-estimate or under-estimate the number of rolls he placed in the proofer? Explain your answer.**

## Managing Costs in Beverage Production

In its simplest, but also its least desired form, beverage production can consist simply of a bartender who **free-pours** drinks.

Free-pouring occurs when a bartender makes a drink by pouring liquor from a bottle without carefully measuring the poured amount. In a situation such as this, it is very difficult to control

**Key Term**

**Free-pour:** Pouring liquor from a bottle without measuring the poured amount.

beverage production costs. At the other end of the control spectrum are automated total bar systems that are extremely sophisticated control devices. Most foodservice operations will be operating under one of the following beverage product control systems:

✓ Free-pour
✓ Jigger pour
✓ Metered pour
✓ Beverage gun
✓ Total bar system

The specific control system utilized will be based on the amount of control operators feel are appropriate for their own beverage businesses.

### Free-Pour

The lack of control resulting from free-pouring alcohol is significant. It should never be allowed when preparing the majority of drinks bartenders will serve. It is appropriate in some settings, however, for example, in wine-by-the-glass sales. In this situation, the wine glass itself serves as a type of product control device. Large operations, however, may even elect to utilize an automated dispensing system for their "wines by the glass." Also, it is most often necessary for a bartender to free-pour when he or she must add extremely small amounts of a product as a single ingredient in a multi-ingredient drink recipe. An example is a bartender who must add a very small amount of dry Vermouth when making a Martini.

### Jigger Pour

A **jigger** is a device used to measure alcoholic beverages that is typically marked in ounces and a fraction of ounce quantities.

Jiggers are inexpensive to buy and use, so this control approach is inexpensive. It is a good system to use in remote serving locations such as a pool area, beach, guest suite, or banquet room. A disadvantage is that there is still room for employee over-pouring and the potential for bartender fraud.

**Key Term**

**Jigger:** A small cup-like bar device used to measure predetermined quantities of alcoholic beverages. These items are usually marked in ounce and portions of an ounce. Examples: one ounce or 1.5 ounces.

### Metered Pour

Some operators elect to control their beverage production costs using a pour spout designed to dispense a predetermined (metered) amount of liquor each time the bottle is inverted.

Pour spouts are inserted into bottles and are available to dispense a variety of different quantities. When using a metered pour spout, the predetermined portion of product is dispensed whenever the bartender is called upon to serve that product.

### Beverage Gun

In some large operations, beverage guns are connected directly to liquor bottles or other containers of various sizes. The gun may be activated by pushing a mechanical or electronic button built into the gun. In this situation, a bartender will, for example, push a gin and tonic button on a gun device, and this will result in dispensing a predetermined amount of both gin and tonic. Although the control features built into gun systems are many, their cost, lack of portability, and possible negative guest reactions can be limiting factors in their selection.

### Total Bar System

The most expensive, but also the most complete, beverage production control solution is a total bar system. This system combines sales information with automated product dispensing information to create a complete revenue and production management system.

Depending on the level of sophistication and cost, the total bar system can perform one or more of the following tasks:

1) Record beverage sale by product brand.
2) Record the individual who made the sale.
3) Record sales value.
4) Measure and dispense liquor for drinks.
5) Add predetermined amounts of mixes to drinks.
6) Reduce liquor values from beverage inventory value totals as drink sales are made.
7) Prepare liquor requisitions.
8) Compute liquor cost by brand sold.
9) Calculate gratuity on checks.
10) Identify payment method (e.g., cash, e-wallet, credit, or debit card).
11) Record guest sales by table or check number.
12) Record date and time of product sales.

## Controlling Cost Through Standardized Recipes

Standardized recipes (see Chapter 1) are essential to any serious effort to produce consistent, high-quality food and beverage products at a known product cost.

All items on a foodservice operation's menu should be based on and follow a standardized recipe. The control of production costs through the management of standardized recipes requires foodservice operators to acquire skills in two key recipe-related areas:

✓ Adjusting standardized recipe yields
✓ Costing standardized recipes

### Adjusting Standardized Recipe Yields

A standardized recipe is a set of instructions used to consistently prepare a known quantity and quality of a menu item. When foodservice operators use standardized recipes, their operations will produce a product that is close to identical in yield and flavor every time it is made regardless of who follows the recipe.

Adjusting standardized recipes (a process also known as **scaling**) means changing the number of servings a recipe produces by multiplying (to increase) or dividing (to decrease) the ingredient quantities in the recipes to match actual production needs.

**Key Term**

**Scaling (recipe):** The process of adjusting the yield of a standardized recipe.

To create an initial standardized recipe, it is always best to begin with a recipe of proven quality. For example, an operator may have a recipe designed to produce 10 portions, but the operator wants to expand it to yield 100 portions. In cases such as this, it may not be possible to simply multiply the amount of each recipe ingredient by 10.

A great deal has been written regarding various techniques used to expand recipes. Computer software designed for that purpose is also readily available to foodservice operators. Generally, any menu item that can be produced in quantity can be standardized in recipe form.

When adjusting standardized recipes to produce a greater or fewer number of portions, it is important that recipe modifications be made properly. For example, weighing ingredients on a pound or an ounce scale is the most accurate method of measuring many ingredients.

It is also important to note that the food item to be measured must be **recipe-ready:** The item must be cleaned, trimmed, cooked, and generally made completely "ready" to be added to the recipe.

For liquid items, measurement of volume (e.g., ounce, cup, quart, liter, or gallon) is usually

**Key Term**

**Recipe-ready (ingredient):** The form a recipe ingredient must be in before it is used in a standardized recipe.

preferred. Some operators and many bakers, however, prefer to weigh all ingredients, even liquids, for improved accuracy.

When adjusting recipes for proper quantity (number of servings desired to be produced), foodservice operators utilize a **recipe conversion factor (RCF)**.

When using the RCF method to adjust a recipe's yield, operators use the following formula to arrive at the appropriate RCF.

**Key Term**

**Recipe conversion factor (RCF):** A mathematical formula that yields a number (factor) operators use to convert a standardized recipe that produces a known yield to the same recipe producing a desired yield.

$$\frac{\text{Desired Recipe Yield}}{\text{Current Recipe Yield}} = \text{Recipe Conversion Factor (RCF)}$$

To illustrate, if a standardized recipe currently yields 50 portions, but the number of portions an operator desires is 125, the RCF formula would be:

$$\frac{125 \text{ desired portions}}{50 \text{ current portions}} = 2.5 \text{ Recipe Conversion Factor (RCF)}$$

In this example, 2.5 is the RCF. To produce 125 portions, the operator would multiply the amount of each ingredient in the original (current) standardized recipe by 2.5 to arrive at the required amount of that ingredient.

Figure 6.2 illustrates the use of this method for a simple three-ingredient recipe.

In some cases, a professionally produced standardized recipe already includes the ingredient amounts needed to make a number of different quantities. For example, a standardized recipe may indicate the ingredients needed to produce 25, 50, 75, and 100 portions of the recipe. In other cases, however, a standardized recipe must be adjusted from its current yield to produce the exact number of portions needed.

Decreasing the yield of a standardized recipe uses the same RCF method as that used when increasing a recipe's yield. To illustrate, assume an operator's standardized

| Ingredient | Original Amount | Recipe Conversion Factor (RCF) | New Amount |
|---|---|---|---|
| A | 4 lb. | 2.5 | 10 lb. |
| B | 1 qt. | 2.5 | 2½ qt. |
| C | 1½ T | 2.5 | 3¾ T |

**Figure 6.2** Recipe Conversion Factor (RCF) Method

recipe yields 80 portions. But the operator wants to make 40 portions. To decrease the recipe yield from 80 portions to 40 portions, the operator would:

1) Calculate the recipe conversion factor (RCF) using the formula:

$$\frac{\text{Desired Recipe Yield}}{\text{Current Recipe Yield}} = \text{Recipe Conversion Factor (RCF)}$$

In this example that would be:

$$\frac{40 \text{ desired portions}}{80 \text{ current portions}} = 0.5 \text{ Recipe Conversion Factor (RCF)}$$

2) Multiply each ingredient amount times the RCF:
    For example, if the original recipe required 12 eggs, and the RCF is 0.5, the number of eggs needed would be 6. In this example, the calculation would be:

$$12 \text{ eggs} \times 0.5 \text{ RCF} = 6 \text{ eggs}$$

One good way to remember the principles of recipe yield conversion is to recognize that, if a recipe's yield is being *increased*, the recipe conversion factor (RCF) will always be *greater than 1.0*. If a recipe's yield is being *decreased*, the RCF will always be *less than 1.0*.

The proper conversion of weights and measurements is important in recipe expansion or reduction. The judgment of the recipe writer is also critical because factors such as cooking time, cooking temperature, and even cooking utensil selection may vary as recipe sizes are increased or decreased. In addition, some recipe ingredients such as spices or flavorings may not respond well to mathematical conversions. In the final analysis, it is an operator's own best assessment of product taste that should ultimately determine proper ingredient ratios in standardized recipes.

---

**Find Out More**

Professional foodservice operators calculate recipe conversion factors (RCFs) for use in increasing and decreasing the size of their standardized recipes. Professional bakers accomplish the same task when they use specialized "baker's math" (sometimes referred to as baker's percentage) ratios to adjust the size of their recipes (or formulas).

When using baker's math, bakers apply the following formula:

$$\frac{\text{Ingredient Weight}}{\text{Flour Weight}} \times 100 = \text{Ingredient Percentage}$$

Essentially, when using baker's math, a baker calculates the total amount of flour called for in a baking formula. Each formula ingredient's weight is then divided by the weight of the flour to determine that ingredient's percentage of the flour's weight).

For example, if a bread formula makes 8 loaves, and it calls for 5 pounds of flour and 1 pound of sugar, the sugar's ingredient percentage would be calculated as

$$\frac{1 \text{ pound sugar}}{5 \text{ pounds flour}} \times 100 = 20\% \text{ sugar}$$

When the formula's sugar's percentage is known, it becomes easy for the baker to adjust the formula's yield for the desired number of loaves.

For example, if the baker wishes to make 12 loaves, the amount of flour required would be 7.5 pounds (5 pounds flour × (12 loaves desired yield/8 loaves current yield) = 5 × 12/8 = 7.5 pounds flour) and the amount of sugar required would be calculated as:

$$7.5 \text{ pounds flour} \times 0.20 \text{ sugar} = 1.5 \text{ pounds of sugar required}$$

It is important to note that, because the weights of the various ingredients in a baker's formula are calculated as a percentage of the flour's weight, the sum of the formula's percentages will always exceed 100%.

To learn more about methods bakers use to adjust their standardize recipes, enter "using baker's math" in your favorite search engine and view the results.

## Technology at Work

Some foodservice operators find it challenging to adjust recipes when they include measurements such as teaspoons, tablespoons, pints, and the like. This can be the case, for example, when an operator's standardized recipe calls for 2/3 teaspoon, and the RFC to be used is 4. In this example, the calculation would be 2/3 t. × 4 RCF = 8/3 or 2 and 2/3 t.

Additionally, while the general U.S. population has been somewhat slow to adopt the international metric system for everyday use, many foodservice operators often find that the standardized recipes they want to use have been developed using metric measurements rather than the imperial (U.S.) measurement system.

When metric measurements are used, standardized recipe adjustment is easier if an operator understands the metric system. Examples: How many pounds are equal to one kilogram? or How many liters are equal to one gallon?

Fortunately, for many conversions, different types of calculators are available online for operators using both the imperial and metric systems. To see how such conversion calculators work, enter "metric conversion calculator" in your favorite search engine and view the results.

## Costing Standardized Recipes

Every menu item sold in a restaurant should be produced from a standardized recipe to ensure product quality and consistency and to assist in product purchasing.

In addition to having a standardized recipe for each item, foodservice operators should maintain a **standardized recipe cost sheet** for each item. In an electronic or hard copy format, operators create standardized recipe cost sheets to calculate their total recipe costs and their recipe's individual portion costs.

**Key Term**

**Standardized recipe cost sheet:** A record of the ingredient costs required to produce a specific standardized recipe.

A standardized recipe cost sheet can be created using any basic spreadsheet software. The spreadsheet (or an operator using a manual calculator) simply multiplies the cost of each ingredient times the amount of the ingredient used in the standardized recipe. For example, if the cost of an ingredient in a recipe is $2.00 per cup and the standardized recipe calls for 3 cups, then the cost for this recipe ingredient is calculated as:

$$\$2.00 \text{ cost per cup} \times 3 \text{ cups required} = \$6.00 \text{ ingredient cost}$$

Computerized spreadsheet programs are an excellent means of creating these records, performing the required mathematical calculations, and keeping recipe costs current. Properly maintained, recipe cost sheets provide operators with up-to-date information that can help with pricing decisions (see Chapter 10) in addition to comparing actual food and beverage costs with those an operation should achieve (see Chapter 12).

A standardized recipe cost sheet can be produced in seconds today using computers and smart device apps. This formerly tedious task has become so simplified there is just no reason for management not to have and use accurate, up-to-date costing data on all of its recipes. As a result, it is easy for managers to know the precise portion costs that should be attainable when their standardized recipes are followed carefully.

To illustrate their use, Figure 6.3 shows the format an operator might use for a standardized recipe cost sheet prepared for a standardized beef stew recipe. The recipe in this example yields a total recipe cost of $43.46 and, because it produces 40 portions, the cost per serving (portion cost) is $1.09 ($43.46 recipe cost/40 portion recipe yield = $1.09 per portion).

Note that all ingredients are listed in their edible portion ("EP") forms, a concept addressed in detail later in this chapter.

Accurately calculating portion costs based on standardized recipe cost is important. However, some operators have difficulty computing total recipe costs when

Menu Item: Beef Stew

Special Notes:

All ingredients weighed as edible

Portion Cost: $1.09

Standardized Recipe Number: 146

Recipe Yield: 40 portions

Portion Size: 8 oz.

Ingredient Cost

| Item | Amount | Unit Cost ($) | Total Cost ($) |
|---|---|---|---|
| Corn, Frozen | 3 lb. | 0.60 lb. | 1.80 |
| Tomatoes | 3 lb. | 1.40 lb. | 4.20 |
| Potatoes | 5 lb. | 0.40 lb. | 2.00 |
| Beef Cubes | 5 lb. | 5.76 lb. | 28.80 |
| Carrots | 2 lb. | 0.36 lb. | 0.72 |
| Water | 2 gal. | N/A | – |
| Salt | 2 T | 0.30 lb. | 0.02 |
| Pepper | 2 t | 12.00 lb. | 0.12 |
| Garlic | 1 clove | 0.80/clove | 0.80 |
| Tomato Juice | 1 qt. | 4.00 gal. | 1.00 |
| Onions | 4 lb. | 1.00 lb. | 4.00 |
| Total Cost | | | 43.46 |

Total Recipe Cost: $43.46

Portion Cost: $1.09

Previous Portion Cost: $1.01

Recipe Type: Soups/Stews

Date Costed: 4/1/20xx

Previous Date Recipe Costed: 1/15/20xx

**Figure 6.3** Sample Standardized Recipe Cost Sheet

recipes contain ingredient amounts that are used in a different quantity than they are purchased.

For example, an operator may purchase soy sauce by the gallon, but their recipes may call for it to be added by the cup or tablespoon. In other cases, ingredients may be purchased in pounds or gallons but are added to recipes in grams and liters. When that is the case, proper conversion calculations must be made.

A second recipe costing challenge often encountered relates to the calculation of product yield (see Chapter 3). Recall that product yield refers to the usable amount of a raw ingredient that remains after it has been cleaned, trimmed, cooked, and portioned. It is an important concept for food buyers, and it is also important when costing standardized recipes.

Most foodservice products are delivered in the as purchased ("AP") state. This refers to the weight or count of a product as a foodservice operator purchases it. For example, if a case of lettuce containing 24 heads is delivered to an operation, the lettuce will be delivered in its "AP" state.

To cost an ingredient for use in a standardized recipe cost sheet, however, the ingredient's edible portion ("EP") cost must be determined. For example, after the 24 heads of lettuce delivered in "AP" condition have been trimmed, washed, and chopped, or otherwise prepared, the heads will now be available in an "EP" condition. Thus, "AP" refers to food products as the operator receives them, and "EP" typically refers to food products as they are used in a recipe or as the guest receives them.

Foodservice buyers purchase ingredients at "AP" unit costs, but these items are used and served in their "EP" forms. To determine actual recipe costs, it is often necessary to conduct a yield test to determine an ingredient's "EP," rather than its "AP" cost.

To illustrate how a yield test results in the determination of actual product costs, assume an operator purchased 10 pounds of fresh carrots ("AP") to be used for stew. The operator knows there will be product loss due to the peeling and trimming of the carrots. As a result, the original 10 pounds of fresh carrots for stew will yield less than 10 pounds when the carrots are peeled and sliced into their "EP" (recipe-ready) state.

The formula used to calculate a product's yield is:

Product As Purchased ("AP") − Losses Due to Preparation = Product Edible Portion ("EP")

Returning to the example of 10 pounds of fresh carrots, if one half pound of carrots are lost in their preparation, the product's "EP" yield would be calculated as:

10 lbs. ("AP") − 0.5 Pounds Loss Due to Preparation = 9.5 lbs. "EP"

Many fresh produce items and many meats and seafoods purchased in their "AP" state will yield significantly reduced amounts when converted to their "EP"

**Key Term**

As purchased ("AP"): The weight, amount, or count of a product as delivered to a foodservice operation.

**Key Term**

Edible portion ("EP") cost: The cost of a product after it has been cleaned, trimmed, cooked, and portioned. Also referred to as a "recipe-ready cost."

**Key Term**

Yield test: A procedure used to determine actual edible portion ("EP") ingredient cost. It is used to help establish actual per usable unit costs for a product that will experience weight or volume loss in preparation.

condition. When performing yield tests, operators must determine a specific product's **waste percentage** and **yield percentage**.

To illustrate the importance of waste percentage and yield percentages to proper recipe costing, assume a foodservice operator purchased 10 pounds of leeks for use in making potato leek soup. After the leeks are cleaned, trimmed, and made recipe ready, 6 pounds of leeks remain. The "EP" yield on the leeks is calculated as:

10 lbs. ("AP") − 4 lbs. Loss Due to
Preparation = 6.0 lbs. "EP"

The leek's waste percentage is calculated using the following formula:

$$\frac{\text{Product Loss}}{\text{"AP" Weight}} = \text{Waste \%}$$

In this example:

$$\frac{4 \text{ lbs. Product Loss}}{10 \text{ lbs. "AP" Weight}} = 0.40 \text{ or } 40\ \%$$

**Key Term**

**Waste percentage:** A ratio obtained by dividing a product's loss amount by its as purchased ("AP") amount.

**Key Term**

**Yield percentage:** A ratio obtained by subtracting an ingredient's waste percentage from 1.00. It refers to the amount of product available for use by an operation after all preparation-related losses have been considered.

In this example, 40% of the leek's weight is lost due to pre-use preparation.

Once a products waste percentage has been determined, it is possible to compute the product's yield percentage using the formula:

1.00 − Waste % = Yield %

In this example:

1.00 − 0.40 = 0.60 or 60%

Yield percentage is important for both buying and recipe costing because, when a product's yield percentage is known, an operator can calculate the "AP" weight to be purchased to obtain the "EP" amount needed in a standardized recipe.

Operators can determine the amount of an "AP" ingredient to buy using the following formula:

$$\frac{\text{"EP" Required}}{\text{Product Yield \%}} = \text{"AP" Amount Required}$$

Foodservice operators can only calculate recipe costs accurately if they know their "EP" (not "AP") costs. The "EP' cost must be known because it represents the true cost of an ingredient or item based on its product yield.

It is important to note that in some cases, the same product may have different yields when purchased from different suppliers. As a result, operators should always use "EP" cost rather than "AP" prices to compare product prices offered from various suppliers. In general, operators want to choose the supplier that offers the lowest "EP" cost for the same product, assuming the same specification or quality is being purchased.

To compute a product's "EP" cost per purchase unit, operators simply divide the "AP" price per pound or purchase unit by the product's yield percentage. In the above leek soup example, with an "AP" price per pound of $1.50 and a product yield of 0.60, the "EP" cost per pound for leeks is $2.50.

This "EP" cost is computed using the "EP" cost formula for items purchased by the pound:

"AP" Price per lb./Product Yield % = "EP" Cost (per lb.)

In this example:

$$\frac{\$1.50 \text{ "AP" Cost per lb.}}{0.60 \text{ Product Yield}} = \$2.50 \text{ "EP" Cost (per lb.)}$$

It is the "EP" cost per pound the operator will use to cost the amount of leeks called for in the operator's standardized recipe for potato leek soup. "EP" costs can and must be calculated for each purchase unit associated with a product (e.g., kilo, crate, bag, box, or carton).

Waste percentage and yield percentage can be known in advance by an operator if good records are kept on meat and seafood cookery. Also, the cleaning and processing of vegetables and fruits must be undertaken correctly to minimize unavoidable losses that can occur in some products during portioning.

Good vendors are often an excellent source for providing information related to trim and loss rates for standard products they sell. With this information, some savvy operators even go so far as to add a minimum or required yield percentage as a part of their product purchasing specifications (see Chapter 3).

---

**Find Out More**

The United States Department of Agriculture (USDA), major food suppliers such as U.S. Foods and Sysco, and a variety of other sources regularly compile and publish typical yield losses for fresh fruits and vegetables, meats, and seafoods.

In some cases, the amount of a product's actual usable yield can be relatively small. For example, the typical yield rate for fresh whole corn kernels cut from whole corn cobs can be as little as 35%. On the other extreme, fresh strawberries may have yields as high as 90% or more.

Regardless of the yield rate of a product, its yield must be accurately calculated before establishing the product's recipe-ready cost. While these food yield values are averages, they can be extremely helpful to foodservice operators as they initially plan food purchases and cost recipes for their own restaurants.

To see some examples of these helpful food yield lists, enter "common food product yields" in your favorite search engine and view the results.

---

**Technology at Work**

A variety of good companies offer software programs that ease the process of standardized recipe development and needed measurement conversion. The reason: Many recipes are produced using metric measurements rather than measurements in the imperial measurement system. Such programs help operators manage their standard recipes files, and they can also:

Calculate total recipe costs
Calculate per portion recipe costs
Help create new recipes
Plan menus
Conduct nutritional analysis of recipes
Track critical allergen information
Monitor product inventory levels in real time

To examine some of these innovative and increasingly essential standardized recipe-related cost control tools, enter "standardized recipe development software" in your favorite search engine and review the results.

## Calculating Actual Cost of Sales

All foodservice operators must be able to calculate their actual cost of sales. It is important to recognize that the cost of sales as shown on an operation's income statement (see Chapter 1) for an accounting period is most often *not* equal to the amount of food and beverage purchases in that same accounting period. In fact, foodservice operators must use a very specific process to accurately calculate their cost of sales and cost of sales percentages for both food and beverages.

### Calculating Actual Cost of Food Sold

Foodservice operators must know their actual cost of food sold during an accounting period. The cost of food sold formula they should use to do so is shown in Figure 6.4.

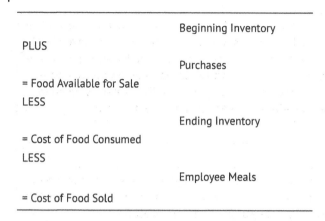

|  | Beginning Inventory |
| --- | --- |
| PLUS |  |
|  | Purchases |
| = Food Available for Sale |  |
| LESS |  |
|  | Ending Inventory |
| = Cost of Food Consumed |  |
| LESS |  |
|  | Employee Meals |
| = Cost of Food Sold |  |

**Figure 6.4** Formula for Cost of Food Sold

While it is commonly referred to as "cost of food sold," or "cost of sales," an operation's cost of food sold is actually the dollar amount of all food sold *plus* the costs of any food that was thrown away, spoiled, wasted, or stolen. To best understand and use the cost of food sold formula properly, operators must fully understand each of its individual parts.

### Beginning Inventory

When calculating the cost of food sold, **beginning inventory** is the monetary value of all food on hand at the beginning of an accounting period.

An operation's beginning inventory can only be accurately determined by completing a physical inventory, which is the actual count and valuation of all foods in storage and in production areas (see Chapter 5).

If, when taking a physical inventory, products are missed or undercounted, an operation's food costs will ultimately appear higher than they are. If, on the other hand, an operator erroneously overstates the value of products in inventory, their costs will appear artificially low.

It is important that operators take accurate physical inventories. For the purpose of calculating food (or beverage) costs, they should recognize that the beginning inventory for an accounting period is always the **ending inventory** amount from the prior accounting period. For example, an

**Key Term**

**Beginning inventory:** The monetary value of all products on hand at the start of an accounting period. Beginning inventory is determined by completing a physical inventory.

**Key Term**

**Ending inventory:** The monetary value of all products on hand at the conclusion of an accounting period.

operation's ending inventory on December 31 will become the operation's beginning inventory for January 1 of the next year.

## Purchases

Purchases are the cost of all food purchased during the accounting period. The purchases amount is determined by properly adding the amounts on all food delivery invoices and any other bills for food products purchased during the accounting period being analyzed. It is important to understand that an invoice need not be paid to be considered part of purchases. Rather, purchases represent the value of all foods accepted for delivery or purchased elsewhere during the accounting period.

## Food Available for Sale

Food available for sale is the sum of the beginning inventory and the purchases made during a specific accounting period. Some operators refer to food available for sale as "goods available for sale" because this term was commonly used prior to the publication of the most recent edition of the Uniform System of Accounts for Restaurants (USAR). Regardless of the term used, both terms represent the value of all food that was available for sale during an accounting period.

## Ending Inventory

Ending inventory refers to the dollar value of all food on hand at the end of the accounting period. It also must be determined by completing an accurate physical inventory.

## Cost of Food Consumed

The cost of food consumed is the actual dollar value of all food used (consumed) by the foodservice operation. Again, note that this amount is not just the value of all food sold. Instead, the number includes the value of all food no longer in the establishment due to sale, spoilage, waste, or theft.

Cost of food consumed also includes the cost of any complimentary meals served to guests as well as the value of any food (meals) eaten by employees.

## Employee Meals

The money spent to provide employee meals (if any) is actually a labor-related, not food-related, cost (see Chapter 8). Free or reduced-cost employee meals are a benefit frequently offered by employers in the same manner as the operation may provide medical insurance or paid vacations. Therefore, the value of this benefit, if provided, should not be recorded as a cost of food. Instead, the dollar value of all food eaten by employees is *subtracted* from the cost of food consumed and then *added* to the cost of labor on the income statement to accurately reflect an operation's actual cost of food sold.

Food products do not have to be consumed as a full meal to be valued as a labor cost. Soft drinks, snacks, and other food items consumed by employees at work are all considered employee meals for the purpose of computing the cost of food sold. If records are kept on the number of employees consuming food each day, monthly employee meal costs are easily determined.

Some operators prefer to assign a fixed dollar value to each employee meal served in an accounting period rather than the actual cost of food eaten by each employee. For example, if an operator assigns a constant value of $6.00 to each employee meal served and 1,000 meals are served in a month, the value of employee meals for that month would be $6,000 ($6.00 per meal × 1,000 meals served = $6,000).

### Cost of Food Sold

An operation's cost of food sold (cost of goods sold) is the actual amount of all food expenses incurred by the operation *minus* the cost of employee meals. It is not possible to accurately determine this number unless a beginning physical inventory has been taken at the start of an accounting period, followed by another physical inventory taken at the end of the same accounting period.

Calculating the actual cost of food sold on a regular basis is important because it is not possible to monitor or improve food cost control efforts unless an operator first knows what the food costs are. In nearly all operations, the cost of food sold is calculated on at least a monthly basis because it is reported on the operation's income statement (P&L). In many operations, these same costs may be calculated on a weekly or even on a daily basis!

It is necessary to properly analyze an operation's cost of sales for a specific accounting period. To do so, operators must first determine the amount of food used in that period and the sales amount achieved in the same period. When they have done so, they can calculate their operation's **food cost percentage**.

**Key Term**

**Food cost percentage:** The portion of food sales that was spent on food expenses.

The formula used to compute an operation's food cost percentage is:

$$\frac{\text{Cost of Food Sold}}{\text{Food Sales}} = \text{Food Cost \%}$$

An operation's food cost percentage represents that portion of food sales that was spent on all food expenses. To illustrate, assume an operation achieved $50,000 in food sales during the current accounting period. In the same period, the operation's cost of food sold was $15,000. The operation's food cost percentage for the period would be calculated as:

$$\frac{\$15,000}{\$50,000} = 30\%$$

Food cost percentage is an important measure for a variety of reasons (see Chapter 12). However, some foodservice operators may be more interested in how much it costs to serve each of their guests rather than their operations' food cost percentages.

Operations in which this average cost of meals served value is important include military bases, hospitals, senior living facilities, school and college foodservice operations, and business organizations that provide free or reduced priced meals to their workers.

Whether the guests served are soldiers, patients, residents, students, or workers, calculating an operation's cost per meal is easy because it uses a variation of the basic food cost percentage formula. The formula used to calculate the average cost of meals served in an operation is:

$$\frac{\text{Cost of Food Sold}}{\text{Total Meals Served}} = \text{Cost Per Meal}$$

To illustrate, assume an operation incurred $35,000 in cost of food sold during an accounting period. In that same accounting period, the operation served 5,000 meals. To calculate this operation's cost per meal, the manager applies the cost per meal formula. In this example:

$$\frac{\$35,000}{\$5,000} = \$7.00 \text{ per meal}$$

Regardless of whether an operator is most interested in calculating food cost percentage or cost per meal served, it is essential that the operator first accurately calculate the amount of their operation's cost of food sold.

## Calculating Actual Cost of Beverage Sold

The proper computation of a beverage cost percentage is identical to that of a food cost percentage with one important difference. Typically, there is no equivalent for employee meals because the consumption of alcoholic beverage products by working employees should be strictly prohibited.

Therefore, "employee alcoholic drinks" would never be considered as a reduction from overall beverage cost. However, it is important to recognize that, like the cost of food sold, an operation's cost of beverage sold is the dollar amount of all alcoholic beverage products sold as well as the costs of all beverages that were given away, wasted, or stolen.

The formula used to calculate actual cost of beverage sold is shown in Figure 6.5.

To properly calculate and analyze the operation's **beverage cost percentage**, its operators

**Key Term**

**Beverage cost percentage:** The portion of beverage sales that was spent on beverage expenses for consumers.

| | |
|---|---|
| | Beginning Beverage Inventory |
| PLUS | |
| | Purchases |
| | = Beverages Available for Sale |
| LESS | |
| | Ending Inventory |
| | = Cost of Beverage Sold |

**Figure 6.5** Formula for Cost of Beverage Sold

must first determine the amount of beverages consumed in an accounting period and the amount of beverage sales achieved in the same period.

The formula used to compute an operation's beverage cost percentage is:

$$\frac{\text{Cost of Beverage Sold}}{\text{Beverage Sales}} = \text{Beverage Cost \%}$$

To illustrate, assume an operation generated $50,000 in beverage sales in the current accounting period. In the same period, the operation spent $9,000 on beverages. The operation's beverage cost percentage for the period would be calculated as:

$$\frac{\$9,000}{\$50,000} = 18\%$$

## Calculating Actual Cost of Sales with Transfers

In some operations, other reductions from, and additions to, food and beverage expenses must be considered when accurately calculating the operation's cost of goods sold. To illustrate, consider the operator of Portillo's, a popular Tex-Mex style restaurant that has a high volume of alcoholic beverage sales.

To prepare drinks, the restaurant's bartenders use a large quantity of limes, lemons, and fruit juices issued from the kitchen. The operator of this business should want their beverage cost percentage to reflect all costs associated with buying and using the alcoholic beverages and the food ingredients normally required to prepare its drinks.

As a result, the operator must transfer the cost of any food ingredients used in drink production from their cost of food sold and add them to their costs of beverage sold.

Assume further that the kitchen uses several standardized recipes that require the use of wines, beers, and liquors issued from the bar. In this case, the operator must transfer the cost of these alcoholic beverages away from the cost of beverage sold and add them to their cost of food sold.

The value of transfers to and from one operating department to another should be recorded on a product transfer form. The form should include a space to record the amount of product transferred, the product's value, and the individuals authorizing the transfers. Figure 6.6 shows how a product transfer form might be used at Portillo's for the first week of January.

In this example, after all appropriate transfers have been recorded for the weeks in the accounting period, the accurate cost of food sold and cost of beverage sold calculations can be made.

| Operation: Portillo's<br>Product Value | | | Month/Date: January 1–7 | | | |
|---|---|---|---|---|---|---|
| Date | Item | Quantity | Cost to Bar ($) | Cost from Bar ($) | Issued by | Received by |
| 1/1 | Lemons | 6 | 3.72 | | T. S. | B. H. |
| | Limes | 2 (large) | 1.28 | | T. S. | B. H. |
| | Cream | 2 qt. | 8.62 | | T. S. | B. H. |
| 1/2 | Chablis | 1 gal. | | 21.10 | B. H. | T. S. |
| 1/3 | Coffee | 2 lb. | 20.70 | | T. S. | B. H. |
| 1/4 | Cherries | ½ gal. | 12.94 | | T. S. | B. H. |
| 1/4 | Lemons | 4 | 3.48 | | T. S. | B. H. |
| | Limes (small) | 12 | 2.24 | | T. S. | B. H. |
| | Ice Cream (vanilla) | 1 gal. | 13.32 | | T. S. | B. H. |
| 1/5 | Pineapple Juice | ½ gal. | 7.00 | | T. S. | B. H. |
| 1/6 | Tomato Juice | 1 case | 20.00 | | T. S. | B. H. |
| 1/6 | Sherry | 750 ml | | 16.70 | B. H. | T. S. |
| 1/7 | Celery | 1 bunch | 0.54 | | T. S. | B. H. |
| Total Product Value | | | 93.84 | 37.80 | | |

**Figure 6.6** Portillo's Weekly Transfer Record

Figure 6.7 shows an example of how the cost of sales formulas are modified for the proper calculation of cost of food sold and cost of beverage sold at Portillo's including transfers for the *entire month* of January.

| | |
|---|---|
| **Accounting Period:** | **20xx: 1/1-1/31** |
| **Operation Name:** | **Portillo's** |
| **Cost of Food Sold** | |
| Beginning Inventory | $ 25,000 |
| PLUS | |
| Purchases | 40,000 |
| = Food Available for Sale | 65,000 |
| LESS | |
| Ending Inventory | (27,000) |
| = Cost of Food Consumed | 38,000 |
| LESS | |
| Food Transfers from Kitchen | (2,000) |
| PLUS | |
| Beverage Transfers to Kitchen | 1,000 |
| LESS | |
| Employee Meals | (1,000) |
| **= Cost of Food Sold** | **$36,000** |
| | |
| **Cost of Beverage Sold** | |
| Beginning Beverage Inventory | $ 12,000 |
| PLUS | |
| Purchases | 18,000 |
| = Beverages Available for Sale | 30,000 |
| LESS | |
| Ending Inventory | (14,000) |
| = Cost of Beverage Consumed | 16,000 |
| PLUS | |
| Food Transfers from Kitchen | 2,000 |
| LESS | |
| Beverage Transfers to Kitchen | (1,000) |
| **= Cost of Beverage Sold** | **$17,000** |

**Figure 6.7** Cost of Food Sold and Cost of Beverage Sold with Transfers

The total cost of sales including transfer amounts shown in Figure 6.7 will be used by Portillo's operator to properly calculate the operation's actual cost of food and cost of alcoholic beverage percentages for the month of January.

Controlling product costs in both food and beverage production is important to all foodservice operators, but their product cost control efforts do not end there. They must also be extremely concerned about controlling costs during customer service. This is true regardless of whether an operator serves most products on-site, through a drive-through window, with customer carry out, or by third-party delivery. Because they are important, controlling product and labor costs during service is the topic of the next chapter.

---

**What Would You Do? 6.2**

"I'm just saying that it seems like a big waste of time to me," said Robin Christopher, the kitchen manager at the Roadhouse restaurant.

Robin was talking to Kyle Minoge, the restaurant's manager.

"Why do you say that?" asked Kyle.

"Well, I just inputted all of the last two weeks' invoice prices into our standardized recipe cost program to calculate new portion costs for all our menu items. It took me almost two hours!" said Robin.

"OK," said Kyle, "what's wrong with that?"

"Well, I just did it two weeks ago and the prices we pay for ingredients haven't really changed that much, and that means the portion costs for the menu items we sell haven't changed that much either!" said Robin.

"Well," replied Kyle, "the owners want us to update our standardized cost sheets every two weeks. That doesn't seem excessive to me. We need to know how much the items we're selling costs us to make."

"Yes," replied a frustrated Robin, "but my point is our costs don't change very much every two weeks, we should just do this once a year!"

**Assume you were one of the owners of the Roadhouse restaurant. What would you think about Robin's suggestion that your standardized recipes cost sheets be updated only once a year? What would be some specific pros and cons about implementing Robin's annual cost update plan when compared to the current bi-weekly update plan?**

## Key Terms

| | | |
|---|---|---|
| Portion cost | Standardized recipe | Yield percentage |
| Free-pour |   cost sheet | Beginning inventory |
| Jigger | As purchased ("AP") | Ending inventory |
| Scaling (recipe) | Edible portion | Food cost percentage |
| Recipe-ready (ingredient) |   ("EP") cost | Beverage cost percentage |
| Recipe conversion | Yield test | |
|   factor (RCF) | Waste percentage | |

---

### Operator's 10-Point Tactics for Success Checklist

Evaluate your need for, and the current status of, each of the following operational tactics. For those tactics you think are important, but not yet in place, develop an action plan for its implementation including who will be responsible for the tactic's completion and the target date by which it should be completed.

| | | | | If Not Done | |
|---|---|---|---|---|---|
| **Tactic** | **Don't Agree (Not Done)** | **Agree (Done)** | **Agree (Not Done)** | **Who Is Responsible?** | **Target Completion Date** |
| 1) Operator recognizes the importance of controlling costs during food and beverage production. | —— | —— | —— | | |
| 2) Operator can calculate the recipe conversion factor (RCF) required to properly increase or decrease the yield of a standardized recipe. | —— | —— | —— | | |
| 3) Operator understands the importance of maintaining an up-to-date standardized recipe cost sheet for each menu item sold. | —— | —— | —— | | |
| 4) Operator can calculate a product's waste percentage when loss due to the product's cleaning, trimming, cooking, and portioning are known. | —— | —— | —— | | |

| Tactic | Don't Agree (Not Done) | Agree (Done) | Agree (Not Done) | If Not Done | |
|---|---|---|---|---|---|
| | | | | Who Is Responsible? | Target Completion Date |
| 5) Operator can calculate a product's yield percentage when the product's waste percentage is known. | —— | —— | —— | | |
| 6) Operator can calculate the as purchased ("AP") amount of an ingredient needed to buy when the ingredient's yield percentage is known. | —— | —— | —— | | |
| 7) Operator can calculate an ingredient's recipe-ready edible portion ("EP") cost per unit. | —— | —— | —— | | |
| 8) Operator can accurately calculate both their cost of food sold and food cost percentage. | —— | —— | —— | | |
| 9) Operator can accurately calculate both their cost of beverage sold and beverage cost percentage. | —— | —— | —— | | |
| 10) Operator can calculate aectual food and beverage costs and their percentages when product transfer costs are known. | —— | —— | —— | | |

# 7

# Cost Control in Food and Beverage Service

**What You Will Learn**

1) The Importance of Cost Control in Food and Beverage Service
2) How to Control Service Costs
3) How to Manage Service Recovery Costs

**Operator's Brief**

Effective control of service-related costs is just as important as the control of product costs. In this chapter, you will learn that, in many ways, the challenges of professional service delivery are even greater than those of product production. This is true, in part, because of the unique service characteristics of intangibility, inseparability, consistency, and limited capacity.

Controlling service-related costs can be challenging because, in many cases, a guest's view of quality service often differs from that of a foodservice operator. In this chapter, you will learn about these differences and their impact on your cost control efforts.

It is important to recognize that the Uniform System of Accounts for Restaurants (USAR) does not have a designated expense category titled "Service." Rather, these costs are recorded in a variety of different expense categories. As a result, foodservice operators controlling their service-related costs often find that these costs vary in large part based on the method used to provide guest service. In this chapter, you will learn about the identification and control of service-related costs for operators providing:

✓ Indoor Dining Service
✓ Outdoor Dining Service
✓ Drive-thru Delivery Service

✓ Guest Pick-up Service
✓ Operator Direct Delivery
✓ Third-party Delivery

Despite your best efforts, your operation will likely experience some service deficiencies. Then you must undertake appropriate service recovery efforts to regain guest satisfaction. These efforts must be addressed for those guests who are on-premises and for those guests who are no longer on-site.

Today, one of the most significant areas of addressing complaints of guests who are no longer on-site involves responding to negative guest reviews posted online. In this chapter, you will learn how to monitor and reply professionally to these negative reviews.

---

**CHAPTER OUTLINE**

The Importance of Cost Control in Food and Beverage Service
    Challenges in Professional Service Delivery
    The Guest's View of Quality Service
    The Operator's View of Quality Service
Controlling Service Costs
    Controlling Indoor Dining Service Costs
    Controlling Outdoor Dining Service Costs
    Controlling Drive-thru Delivery Service Costs
    Controlling Guest Pick-up Service Costs
    Controlling Operator Direct Delivery Service Costs
    Controlling Third-party Delivery Service Costs
Managing Service Recovery and Recovery Costs
    On-Premises Service Recovery
    Off-Premises Service Recovery

---

# The Importance of Cost Control in Food and Beverage Service

To best understand the importance of controlling costs during service, it is essential to recognize that a foodservice operator is both a manufacturer and a retailer. A professional foodservice operator is unique because all functions of a product's sale, from menu development to guest service, are normally the responsibility of the same person.

Foodservice operators must secure raw materials (menu ingredients), produce a product, and sell—often under the same roof. Few other managers require the breadth of skills required to be effective foodservice operators. Since operators are in the service sector of business, many aspects of management are more challenging for them than for their manufacturing or retailing management counterparts.

A foodservice operator is one of the few types of managers who have contact with the ultimate consumer. This is not true, for example, for the managers of a cell phone factory or automobile production line. These individuals produce a product, but they do not sell it to the person who will use it. In a like manner, furniture and clothing store managers sell products to those who use them, but they do not have a role in producing the products they sell.

The face-to-face guest contact in the foodservice industry requires operators to assume the responsibility of standing behind their own work and that of their staff. This often occurs in a one-on-one situation with the ultimate consumer (end user) of an operation's products and services.

The management task checklist in Figure 7.1 shows some areas in which foodservice operators, manufacturing managers, and retailing managers differ in their responsibilities.

In addition to their role as food factory supervisors, operators must also serve as a cost control manager during service, because, if they fail to perform this vital role, their business will perform poorly or may even cease to exist.

Unlike product costs that are fairly easy to count and measure, service-related costs in a foodservice operation exist in a number of expense categories, but typically will fall into one of three broad and sometimes overlapping groupings, each of which must be controlled. These are:

1) Labor-related costs
2) Product presentation-related costs
3) Product delivery-related costs

### Labor-related Costs

Labor-related service costs consist primarily of wages and benefits paid to front-of-house staff who are in direct contact with guests including hosts, servers,

| Task | Foodservice Operator | Manufacturing Manager | Retail Manager |
|---|---|---|---|
| 1) Secure raw materials | Yes | Yes | No |
| 2) Manufacture product | Yes | Yes | No |
| 3) Market to end user | Yes | No | Yes |
| 4) Sell to end user | Yes | No | Yes |
| 5) Reconcile problems with end users | Yes | No | Yes |

**Figure 7.1** Management Task Checklist

bussers, and bartenders. The costs incurred to employ these staff can be significant and, as a result, careful staff scheduling and appropriate training to optimize worker productivity is essential.

### Product Presentation-related Costs

Product presentation-related costs are those that primarily focus on visual or sensory appeal. Some examples of these costs include employee uniforms, silverware, glassware, delivery items (if provided), and napkins. Additional examples include artwork and decor as well as live or overhead music.

### Product Delivery-related Costs

This service-related cost area is of increasing importance to many foodservice operators as the demand for off-premises dining by guests continues to increase. Product delivery-related costs can include menu item packaging for off-site dining, transportation costs (if products are delivered directly by an operation to its guests), and any delivery/commission fees paid to **third-party delivery** partners selected by the operation.

**Key Term**

**Third-party delivery:** The use of a smartphone or computer application that allows customers to browse restaurant menus, place orders, and have them delivered to the customer's location. In nearly all cases, the requested orders are delivered by independent contractors retained by the company operating the third-party delivery app.

### Challenges in Professional Service Delivery

Foodservice operators sell products, and they deliver services. In most cases, it is easier for foodservice operators to control quality in the production process than it is to control quality in service delivery. As a result, it is not surprising to know that guest complaints about service delivery quality in a foodservice operation typically far exceed those complaints related to product quality.

For example, guests who order a 14-ounce ribeye steak know it will always be larger than a 10-ounce ribeye. Priced properly, guests may perceive better value in the larger steak (such as when the selling price per ounce is less for the larger steak). Similarly, French fries may be offered in a small or large portion size. In both the above examples, buyers can easily be told what they will be getting for the prices they will be charged. If production quality is controlled properly, high levels of guest satisfaction are the likely result.

When a foodservice operation advertises that it offers quality service, good service, or quick service, however, these concepts may not be so easy to communicate and deliver. Consider, for example, that an operation's service quality is affected by several unique characteristics of service including:

✓ Intangibility
✓ Inseparability
✓ Consistency
✓ Limited capacity

### Intangibility

Foodservice customers buy products, but they receive services, and the sale of a service provides an intangible benefit to a foodservice guest. It is intangible because, unlike a physical product, a service cannot be seen, tasted, felt, heard, or smelled before its purchase. In most cases, service is more of a performance than a product. Since they are intangible, operators face unique challenges in communicating the benefits of services offered to those who will buy them.

Guests buying from a foodservice operation must put their faith in the operator that the service provided will be of high quality. Foodservice operators must then justify this trust by consistently delivering service at a level that meets, and even exceeds, their guests' expectations.

### Inseparability

Inseparability refers to the idea that it is often not possible to separate the production of a service from a guest's consumption of the service. Foodservice guests purchase and use a service at the same time. In a customer service business, guest satisfaction often depends on how the guests interact with employees during their visit.

A table service guest who judges their server to be exceptionally helpful will most often attribute positive aspects to the foodservice operation itself. Inseparability is an important concept because it helps operators to consider many factors that contribute to their success. These include guests who most often equate the quality of service they receive with the actual person who provides it. Therefore, the appearance of a server's uniform, the friendliness and facial expressions exhibited during service, and even the words used to communicate with guests can affect the guests' perceptions of service quality.

Experienced foodservice operators know, for example, that a guest could post a negative review of their operation online even if the food was good. This can be the case when the food is excellent, but the service received by the guest was not good. Alternatively, if an operation's food was only average, but the service received by the guest was exceptional, the guest may overlook the quality of the food and still be left with a very positive perception of the operation.

## Consistency

Consistency in service can be a challenge to achieve because the quality of a service often depends on the individual who supplies it. Inconsistency when providing services is usually much greater than when providing products. For example, the customers at a sports bar will usually find that the quality of several bottled beers of the same brand purchased while watching a game is identical. However, the skill level, appearance, amount of attention provided, and attitude of the beer's server can vary greatly, even during the same game.

In this example, these potential differences in product delivery can have a direct effect on the quality of the customer's beer purchase experience. Due to the importance of consistency when providing hospitality goods and services, it follows that foodservice operators should pay a great deal of attention to their operations' front-of-house staff training and service standardization efforts and activities.

## Limited Capacity

Nearly all foodservice operations face limitations in their capacity. In the front of the house (public areas), capacity may be limited by the number of seats in the dining room and/or the number of drive-thru lanes and service windows available.

In the back of the house (employee-only areas), capacity may be limited by the number of workers available or the production capability of an operation's preparation equipment. As a result, it may be easier to deliver high-quality service during times of lower ("average") capacity than at times of full capacity. Foodservice operators seek to attract sufficient customers to operate at near full capacity because, unlike product inventory that can be held over from one day to the next, lost revenue from a seat that remained empty during a service period is lost forever.

Since most foodservice operations face limited capacity, it is especially important that their managers assign service staff based on anticipated guest demand levels. Too many workers scheduled during slow times may cause an operation's labor-related costs to rise to unacceptable levels. However, too few workers available during busy times may cause an operation's service levels to suffer, even when its service workers are well-trained and are doing their very best to serve guests.

Despite the challenges often encountered during the delivery of high-quality service, foodservice operators must ensure that guest service levels are high and service-related costs are well managed. These results are integral to efforts that help to achieve a foodservice operation's reputation for providing both excellent products and excellent service.

## The Guest's View of Quality Service

Experienced foodservice operators know a guest's view of service quality is impacted each time the guest experiences a **moment of truth:** any point in time where guests draw their own conclusions about the levels of a foodservice operation's service or product quality.

> **Key Term**
>
> **Moment of truth:** A point of interaction between a guest and a foodservice operation that enables the guest to form an impression about the business.

Foodservice operators recognize that moments of truth can occur even before a foodservice guest arrives at an operation. For example, if a foodservice operation's marketing efforts result in a potential guest visiting the operation's website to place an online order, then the website must be easy to use and functional. If it is not, guests will likely conclude that the operation is not service-oriented. In this digital marketing example, the guest's moment of truth is most impacted by the actions of management when it developed the operation's website and not the operation's actual ability to deliver high-quality menu items in a timely manner.

If foodservice operations offer on-premises table service, the service staff personally manage moments of truth as guests interact directly with them. While each foodservice operation is unique, guests in many operations will assess service quality as an operation's staff undertake key activities in the following areas:

1) Greeting guests
2) Taking guests' orders
3) Delivering guests' orders
4) Serving guests as they enjoy their meals
5) Collecting payment

The effective management of each of these key steps is critical if an operation's staff is to deliver high-quality service.

### Greeting Guests

A friendly and professional initial greeting for arriving guests is important. However, foodservice operators should recognize that a guest's first moments of truth can actually occur well before they enter an operation's dining area or its carry out pick-up station.

For example, guests may form an initial opinion of an operation when they visit the operation's website, arrive in the operation's parking lot, and approach the operation's main entrance. In each of these cases, guests can assess the service quality of the operation even before they are seated in the dining room or receive their "take home" meal.

The physical appearance of a facility, including clean, well-lighted parking and walk-up areas, helps to form positive guest perceptions about an operation's quality and about its attention to detail. The physical appearance of an operation gives guests a good idea about what they can expect to find inside the foodservice operation. Moments of truth occur each time guests visually inspect the cleanliness of an operation's windows, its foyer and waiting areas, and even its guest restrooms.

Additional visual clues guests will encounter relate to the appearance of staff and their uniforms. Regardless of the level of formality in the operation, employee uniforms should be clean, fit properly, and include an easy-to-read name tag. Guests who observe employees in dirty, wrinkled, or inappropriate uniforms will likely draw negative conclusions about the quality of food and service they will receive. For that reason, an operation's policies on issues such as hair color and styles, visible tattoos, and body piercings must make sense and fit with the operation's overall image.

Whether they arrive at an upscale operation's entrance foyer or at a drive-thru window, each guest's initial greeting by an operation's staff should reflect a friendly and hospitable attitude. When that occurs, guests will likely form a positive opinion of the operation. If it does not occur, guests will notice, and the operation will have to work very hard to recover to overcome guests' initial and negative moments of truth.

### Taking Guests' Orders

While the actual method used to take a guest order varies based on the type of operation, it is important for operators to know that proper order taking can be a skillful art. When staff have been trained to take orders politely and to provide guests with accurate information regarding questions they may have about the menu, positive moments of truth will occur.

Inaccurate order taking results in guests who do not receive the items they had ordered and/or planned to receive. When order taking is performed improperly, an operation's kitchen staff may prepare ordered items perfectly, yet guests will still be disappointed in the items they receive.

### Delivering Guests' Orders

Today's foodservice guests can obtain their orders in numerous ways. The traditional method of delivering guest orders to those seated at tables is still prevalent in many foodservice operations. When guest orders are delivered this way, it is important that service staff "get it right." Proper server training programs help ensure that all servers use a system to ensure the right order is delivered to the right guest.

The proper delivery of guest orders is even more important when guest pick-up and drive-thru delivery systems are used. This is true because, with these two delivery methods, guests will not actually see their individual menu items when they are received. Rather, the menu items will likely be packaged in ways that may make it difficult for guests to know if they have received the correct order until they return home or to another location. Then, it is likely too late to correct any order errors, and guests will experience a negative moment of truth.

In increasing numbers of cases, the delivery of guest orders may involve the use of a third-party delivery company. However, operators must understand that moments of truth for a guest occur regardless of whether their orders are delivered by a third-party or by the operation itself.

### Serving Guests as They Enjoy Meals

Operators are aware that well-trained service staff can be a significant determinant when table service is used because service extends beyond providing the initial order. Examples may include the need for a manager to "check back" to confirm all is well with the guests' orders, to determine if a guest wants a beverage order (including water, soft drink, or coffee), and, perhaps, to sell a dessert or after-meal beverage.

The best servers do not normally ask, "Is everything okay?" Instead, a more friendly example, "I hope everyone is enjoying their meal. May I bring you anything else?" can inform guests that the server is available to provide additional assistance.

Experienced operators know the importance of a well-trained service team that works together. For example, when refreshing coffee or water, servers can assist more than just the guests in their own work sections. If a problem is noted, the appropriate server can be informed about assistance that is needed.

### Collecting Payment

An operation's guest payment system has the potential to produce positive or negative moments of truth. For those operations providing on-site dining, a negative moment of truth can occur when guests must wait excessively long to receive their checks, or when presented checks take an excessive time to be processed for payment. Alternatively, if guest checks are presented too quickly, on-site diners may feel they are being pressured to leave the restaurant. Well-trained servers can anticipate their guests' preferences and present their checks to them at the appropriate time.

Even when delivered at the proper time, a guest check that contains errors (e.g., when guests are charged for items not ordered, or incorrect prices are charged for ordered items), negative moments of truth are the inevitable result.

| Technology at Work |
| --- |
| Successful marketing programs can attract new guests to a foodservice operation. However, the ways in which on-site staff manage their guests' critical moments of truth are often determining factors that impact the guests' return to the foodservice operation. |

Successful marketing programs can attract new guests to a foodservice operation. However, the ways in which on-site staff manage their guests' critical moments of truth are often determining factors that impact the guests' return to the foodservice operation.

One effective way to help ensure that staff have the skills necessary to successfully serve guests is to identify and implement an effective guest service training program.

All foodservice operations, from quick service restaurants to fine dining establishments, must provide an excellent experience for their guests. This experience includes everything from taking guests' orders to delivering selected menu items appropriately.

High-quality guest service training can improve customer skills for all employees who interact directly with guests. The good news is that online customer service training tools can be obtained and utilized very inexpensively or even at no cost.

To review some currently popular free-to-use guest service training programs for foodservice operations, enter "free service training for restaurant staff" in your favorite search engine and review the results.

Increasingly, foodservice guests assess the quality of an operation's service not through the moments of truth they experience, but rather through the experience of others. Potential guests read about, and even view, the experiences of others as they search **user-generated content (UGC) sites** for information about an operation's service (and product) quality as described by an operation's previous guests.

UGC sites of most interest to foodservice operators are those that focus entirely, or at least primarily, on guest reviews. Currently popular sites include Yelp, Open Table, TripAdvisor, Zomato (formerly Urbanspoon), and more. Using stars or

**Key Term**

**User-generated content (UGC) site:** A website in which content including images, videos, text, and/or audio have been posted online by the site's visitors.

Examples of currently popular UGC sites include Instagram, X (formerly Twitter), Facebook, TikTok, Pinterest, and YouTube.

points (typically on a scale of one to four or one to five), these sites allow guests to rate the quality of service and food they have received during a dining experience, and to then share their ratings online. Guests searching for information about a specific foodservice operation will increasingly go online to review the average "point" or "star" value assigned to that operation by previous guests.

Some foodservice operators are wary of reviews guests post on social media and UGC review sites, and a cautious reaction is not surprising. In many cases, those who operate foodservice facilities deal directly with guests when a problem arises. If, for example, an operator serves 100 guests in one day, and five customers have a complaint, the operator might feel that "many" customers are dissatisfied. In this example, however, 95% of this operator's customers did not encounter a problem worth mentioning.

It is important to recognize that in many cases online reviewers are more motivated to post *positive* reviews about foodservice operations rather than negative reviews. While consumers do share experiences when they are dissatisfied, they are much *more likely* to share experiences that show their selection of a foodservice operation was a good one.

Professional foodservice operators recognize the power of popular UGC review sites because they know that people more readily trust the recommendation of a real person over what a business says about itself. In a recent study, 56% of consumers said they more readily trust user-generated content, and that it was what they most wanted to see from businesses. Note: Only 15% of the consumers said content created by businesses was what they wanted to see.[1]

Foodservice guests typically feel compelled to submit ratings and reviews to a UGC site if they:

✓ Have a positive experience
✓ Are genuinely wanting to help the operation improve its products and/or services
✓ Are motivated to give a special "thanks" to one or more individual staff members
✓ Have been incentivized in some way to do so
✓ Have a negative experience
✓ Want to contribute to the social media community of which they are a member
✓ Want to encourage (or warn!) other site users about a specific foodservice operation

Recognizing the importance of ratings and user reviews to their online reputations, foodservice operators should continually monitor popular UGC review sites to see how their own guests have reviewed and rated their facilities for service and product quality. Note: Detailed information about how best to respond to negative UGC reviews is presented later in this chapter.

---

### What Would You Do? 7.1

"We've got to fix this," said Cindy, the dining room manager at the Buttercup Bistro.

"Fix what?" asked Marion, the Bistro's kitchen manager.

"This average score we're getting on the 'DineInTown' website," replied Cindy.

The "DineInTown" website featured guest reviews of local restaurants, and it was a popular site with many reviewer postings. The site allowed reviewers to

---

1 https://www.nosto.com/blog/what-is-user-generated-content/, retrieved September 1, 2022.

rank restaurants on a scale of 1 to 5 (with 1 being a low score, and 5 being the highest score).

"We are averaging a score of 3.8," said Cindy. "I don't expect us to get all 5s, but we should at least be in the 4 plus range. The interesting thing is that we tend to get almost all 5s because reviewers say they love our food, and reviewers who give us a score of 1 almost always refer to slow service. These are mostly guests who visited us on a Friday or Saturday night. With lots of 5s and then a lower score mixed in, we end up with a 3.8 average. That's what we need to fix."

**Assume you were Cindy. What do you think your guest reviews are telling this operation about its service staffing patterns on Friday and Saturday nights? How important would it be for you to carefully manage reservation capacity as well as service and production staffing levels on these two busy nights?**

## The Operator's View of Quality Service

All foodservice operators want to provide good service to their guests, and most operators feel that they do so. These operators tend to view quality service not in terms of moments of truth, or even guest reviews, but rather they are concerned about three key service-related factors:

✓ Speed
✓ Accuracy
✓ Professionalism

### Speed

In many foodservice operations, management believes their guests will most often equate quality of service with speed of service. These managers recognize that guests are not likely, for example, to complain that their drive-thru orders were prepared too quickly! Guests will, however, complain if drive-thru lanes are long and/or if drive-thru order delivery times are excessive.

Similarly, in an on-premises dining table service restaurant, guests will not likely complain about being seated or served too quickly. There may be complaints, however, if guest wait times are excessively long. This is especially true if guests have made a reservation in advance and are not seated close to their reservation time.

Knowledgeable operators know that monitoring drive-thru times for those operations providing this service and **ticket times** for those operations providing on-premises dining are good ways to continually monitor speed of service. Shorter service times are preferable to longer service times, and desired standards for these times should be established and continually monitored.

**Key Term**

**Ticket time:** The amount of time required to fill a guest's order or respond to a guest's request.

The actual ticket time standards that are appropriate for an individual operation vary based on the menu items it sells and the service levels offered. However, some generally acceptable time-related standards common in the foodservice industry include:

✓ On-premises dining:
- Guest greeting and seating: 2 to 5 minutes from arrival
- Drink service: 1 to 3 minutes from seating; 4 to 6 minutes from ordering
- Appetizer service: 5 to 10 minutes from ordering
- Entrée service:
  ○ Lunch: 8 to 12 minutes from ordering
  ○ Dinner: 12 to 15 minutes from ordering
- Dessert service: 3 to 5 minutes from ordering

✓ Drive-thru service: 4 to 6 minutes from order placement to order delivery

✓ Carry out (pick-up) service: 15 to 30 minutes from order placement to order pickup

✓ Third-party delivery: 30 to 45 minutes from order placement to order delivery

---

**Find Out More**

Foodservice guests do not like to wait an excessively long time to be served. This is especially true for guests who are utilizing a foodservice operation's drive-thru service.

Interestingly, drive-thru times in foodservice operations have been increasing in recent years rather than declining. Part of this challenge occurs because quick-service restaurants continually expand their menus as they seek to attract more guests. While this can be a positive thing, it can also increase drive-thru times as guests may take longer to choose their items from an expanded list of alternatives.

Average drive-thru times for foodservice operations are regularly measured and publicized. The variation in drive-thru ticket times can be significant. For example, a recent survey found that drive-thru times at Taco Bell averaged 221.99 seconds (the shortest ticket time of the operations surveyed) while Chick-fil-A averaged 325.47 seconds (the longest ticket time of the operations surveyed).[*]

To monitor changes in average ticket times in the foodservice industry and to gain insight on current strategies to reduce these times, enter "tips for reducing drive-thru ticket times in restaurants" in your favorite search engine and view the results.

[*] https://www.delish.com/food-news/a41648070/restaurants-drive-thru-wait-times/, retrieved February 28, 2023.

## Accuracy

From an operator's cost control perspective, accuracy is second only to speed in helping to ensure guest satisfaction. Certainly, when guests do not receive the menu items they have ordered, or when the prices charged for those items are incorrect, guest satisfaction levels will be reduced.

In addition, when the wrong menu items are produced and presented to guests, and those items are not accepted because they are not what the guest ordered, product-related costs rise due to product waste. This is because, in most cases, menu items that typically are incorrectly prepared and delivered to a guest cannot be served to other guests.

For example, if a guest requests a dinner salad "with Ranch dressing on the side," but in fact the salad is served with the dressing already added to the salad, that salad could probably not be served to another guest. In this example, salad product costs will be doubled due to improper accuracy in filling the guest's order.

To help minimize guest dissatisfaction due to inaccuracy of orders, servers must be carefully trained in order taking. In many cases, it is a good idea to repeat a guest's order fully before submitting the order to the operation's production staff for item preparation.

## Professionalism

Experienced foodservice operators know that exhibiting professionalism in service delivery constitutes a variety of factors. These include providing facilities that are clean, properly lit, and held at a temperature that permits comfortable dining.

Staff-related factors that impact guests' views of an operation's professionalism include the wearing of clean uniforms, practicing appropriate personal hygiene, and projecting an image that reflects well on the operation.

When servers are polite, friendly, and demonstrate genuine concern for guest satisfaction, an operation will project the type of professionalism that guests expect and that will lead them to believe service levels are good. If order takers or servers are rude, appear rushed, or seem unconcerned about guest satisfaction, guests will respond by feeling that service levels were poor, even if speed and accuracy of service were acceptable.

# Controlling Service Costs

In many cases, foodservice operators addressing their service-related costs will focus the greatest amount of attention on labor-related costs. It is true that front-of-house labor costs for positions such as hosts, servers, order takers, bussers, and the like will constitute a significant portion of an operation's service costs.

The control of front-of-house staff costs (and **back-of-house staff** costs) is important and will be addressed in detail in Chapter 8. Other service-related costs, however, must be controlled as well.

It is important to recognize that the USAR format for an income statement does not include a specific line for "service-related" costs. Rather, these costs will be listed in a variety of designated cost categories. For example, the cost of live music provided to entertain guests while they are dining would be listed in "Music and Entertainment." The cost of providing controlled temperature conducive to guest dining will be listed under "Utilities."

**Key Term**

**Back-of-house staff:** The employees of a foodservice operation whose duties do not routinely put them in direct contact with guests.

As a result, one good way to review specific areas in which control efforts must be directed is to consider how an operation delivers its service. Each delivery style will have unique service-related costs to be assessed and managed. In most cases, foodservice operators will be concerned with controlling specific types of service costs when they provide their guests with one or more of the following types of meal service:

✓ Indoor Dining
✓ Outdoor Dining
✓ Drive-thru Delivery
✓ Guest Pick-up
✓ Operator Direct Delivery
✓ Third-party Delivery

## Controlling Indoor Dining Service Costs

The control of service-related costs when providing indoor dining to guests include those expenses related to the maintenance of an appropriate **ambience,** and the environment created by an operation's heating, ventilating, and air conditioning (HVAC) system, as well as its lighting costs.

**Key Term**

**Ambience:** The character and atmosphere of a foodservice operation.

Additional indoor dining costs that must be controlled include the costs of linen and napkins when those items are in use, as well as flatware, glassware, and china. In most indoor dining restaurants, the cost of employee uniforms will also be significant, and these must also be properly managed and controlled.

## Controlling Outdoor Dining Service Costs

Outdoor dining, also known as "alfresco dining" or "dining alfresco," is eating outside. Unique service-related costs incurred when an operation provides outdoor dining for guests can include the cost of operating appropriate HVAC units and providing appropriate lighting. Entertainment-related costs such as live music are also common when providing outdoor dining services.

As part of the shift to increasingly popular outdoor dining, some foodservice operators now provide outdoor **dining bubbles**. Also known as bubble tents or dining pods, these enclosed and controlled environments allow guests to be served in colder (or hotter) weather and can be placed on sidewalks or in parking areas.

**Key Term**

**Dining bubble:** A controlled environment used to house foodservice guests while they are dining outdoors.

Typical types of dining bubbles include zip-up tents that can house tables for up to four or six diners at a time, and large see-through geodesic domes of various sizes. The price for dining bubbles varies based on their size and construction quality.

While every outdoor dining facility is unique, it is important to recognize that operators providing such services must comply with the Americans with Disabilities Act, and this requires ramps and spaces to allow mobility aids to move freely. In addition, in some locations offering outdoor dining, the cost of elimination and prevention of outdoor pests including flies, mosquitoes, birds, and vermin must be considered. Also, in some communities, operators may have to pay an extra permit fee to provide outdoor dining.

## Controlling Drive-thru Delivery Service Costs

Increasing numbers of foodservice operators offer their guests a drive-thru service option. While historically drive-thru service was provided primarily by quick-service restaurant (QSR) operators, today more operations choose to provide this popular service delivery style.

While it is possible (but extremely difficult) to provide drive-thru service without a **digital menu board,** in most cases operators providing drive-thru delivery service incur the cost of creating and maintaining this important equipment.

A digital menu board is a complex system of hardware and software used to display menus on screens read by guests. In addition to the display of menu items and prices, digital menus can also display specials and promotions, pictures, videos, customer reviews, and more.

**Key Term**

**Digital menu board:** An integrated system that uses hardware and software to display an operation's menu on an electronic screen; also commonly referred to as a digital display menu or digital menu board.

Digital menus for drive-thru operations are designed to be viewed from the operation's exterior, and it is the type most often seen in foodservice operations offering drive-thru service. When they are placed outside, they must be weatherproof to shield the hardware from environmental elements. They must also be properly designed to be easily read in bright sunlight and at night. They must also be constructed to minimize their chance of damage from vandals or careless drivers.

Costs related to operating a digital menu sign include speaker systems, display screens, and a media player: the device that downloads menu content from the sign's software. The **content management system (CMS)** in a digital menu board is critical as it allows operators to create the content their guests will view, to upload new content, and to then change or modify content as needed.

The hardware cost for a very simple two-screen (one inside and one outside) digital menu board system ranges from $2,000 to $5,000, with monthly fees for digital signage software and support ranging from $50 to $100 per month.

## Controlling Guest Pick-up Service Costs

In a guest pick-up service style, guests receive their menu items from drive-thru windows or pick up their ordered items from designated on-premises carry out areas. In addition to providing appropriate staffing and space for pickup, foodservice operators offering this service are most concerned with proper menu item packaging and packaging costs.

The packaging of pick-up (take away) items is just as important to ensuring guest satisfaction with the service as is the quality of the menu items being packaged. To illustrate, assume a foodservice operation received an order for a 3 chicken taco meal. The tacos *could* be fully assembled and placed in a Styrofoam container and, if so, the food would technically be ready for pick-up.

It would be much better, however, if the chicken fillings were packaged in a hot container, chilled items like sour cream, cheese, and shredded lettuce were packaged in a cold container, and the warm tortillas were wrapped in aluminum foil. In this second packaging scenario, while packaging costs would be increased, even if the guest's pick-up time were delayed, product quality would be optimized.

While each foodservice operation is different, fundamental "good service" principles of take away food packaging include:

✓ Packaging hot foods separately from cold foods
✓ Packaging liquid foods in containers with secure fitting lids
✓ Utilizing see through containers whenever possible

✓ Replacing plastic containers with more environmentally friendly containers when possible

✓ Customizing take away containers and products with the operation's name or logo where it is economically feasible to do so

It is important for foodservice operators to recognize that, even when menu items are carefully selected for suitability as a takeout item, and if they are packaged properly, product quality can still decline. This is especially so if these items are not held correctly until they are picked up by guests and/or by delivery drivers.

Guest pick-up of ordered items can be delayed by traffic problems, other unforeseen circumstances, or simply a guest's miscalculation of travel time. To minimize the negative impact on food quality of excessive holding times, operators must ensure that hot food packages are held separately from cold food packages. In many cases, this will require incurring the costs of separate containers or bags for a single pick-up order. If hot and cold foods must be packaged together in the same bag, box, or container, the order should be assembled upon guest or driver arrival, whenever possible, to help ensure guest perceptions of high service quality.

## Controlling Operator Direct Delivery Service Costs

When utilizing operator direct delivery, menu items are delivered to guests by an operation's own delivery employees. Direct service-related costs when utilizing this approach include those of driver wages and benefits, vehicle costs, and fuel costs.

For guest orders that will be delivered in coolers or insulated bags, the costs of proper cleanliness, sanitization, and disinfection between uses will be incurred. Increasingly, some operators add tamper-evident labels (those that, when ripped or torn, alert guests that their order was tampered with), or tamper-proof boxes that can also be used for their own deliveries.

While there are certainly service-related costs to self-delivery, many operators point to advantages to implementing a self-delivery program that include:

✓ Quality control of delivered foods

✓ Ability to set their own delivery charges

✓ Use of logoed uniforms and transportation vehicles to raise operation's visibility among guests and potential guests

✓ Increased control of guest data

✓ Elimination of third-party commission and fees

✓ The potential ability to utilize current staff for deliveries during slow production or service periods

---

**Technology at Work**

As takeout meals and third-party delivery companies continue to grow in popularity, the need for appropriate takeout packaging continues to evolve.

Peace of mind for customers receiving delivered meals is important and, as a result, there has been increased demand for tamper-proof, or tamper-evident, packaging of delivered meals. This type of packaging helps ensure that food has not been "sampled" or otherwise adulterated by delivery drivers.

Tamper-proof packaging provides peace of mind both to guests and to foodservice operations. Guests can be assured that their delivered menu items are safe and unadulterated. Foodservice operators can utilize third-party delivery services with reduced fear of tarnishing their operations' name by unscrupulous delivery drivers. Tamper-proof containers typically cost more than those container types that do not offer such features so this is a primary cost concern as well.

Innovative packaging for takeout and delivery foods continues to evolve. To learn more about the tamper-proof products resulting from this evolution, enter "tamperproof packaging for restaurant delivery" in your favorite search engine and review the results.

---

## Controlling Third-party Delivery Service Costs

Third-party delivery costs are incurred when menu items are delivered to guests by one or more third-party delivery partners selected by a foodservice operation. The COVID pandemic that began in 2019 significantly increased guest use of third-party delivery services. During the height of the pandemic, many foodservice operations were closed for indoor dining and, also, many guests decided it was safer to eat in their homes than dine out.

The basic business model of third-party delivery companies is relatively straightforward. Foodservice operators agree to have their operations listed on the app developed and marketed by the third-party delivery company, and the operator then provides a menu that includes item prices. Guests utilizing the apps place their orders with the third-party delivery company (not directly with the operation).

Orders placed on the app are electronically transmitted to the foodservice operation's point-of-sale (POS) system. The third-party delivery company uses a cadre of independent drivers to then pick up the orders and deliver them to guests. In addition to paying for the food they have ordered, customers also pay a delivery fee to the third-party delivery company. Drivers are paid on a per order basis by the company, and they typically retain any tips generated by the orders they have picked up and delivered.

The third-party delivery company charges a commission to the foodservice operation for their delivery services. Typical commission fees range from 20% to 30% of the dollar amount of the order. The remaining amount of the food order's value is then electronically transferred to the foodservice operation's bank account.

Those foodservice operators who partner with third-party delivery services typically do so with the belief their profitability will increase. In fact, third-party delivery companies often indicate that cost savings do accrue to foodservice operations that do not need to create a unique online order and delivery system because it is provided by the third-party service.

Foodservice operators can incorporate online ordering and delivery service into their own websites, but there is a cost to do so. In addition to setting up an online ordering system, operators implementing their own delivery service incur additional costs of operating a delivery fleet and paying its drivers.

Despite their increasing popularity, some foodservice operators seeking to control service costs and ensure their own profitability are concerned about the high costs of third-party delivery partnerships. For foodservice operators, the delivery fees paid to third-party delivery app operators reduce the amount of money that remains to pay for food, labor, and other operating expenses required to produce the menu items that are delivered.

To illustrate the financial impact of partnering with a third-party delivery app, consider the data presented in Figure 7.2, an income statement (see Chapter 1) for the El Chico restaurant prepared using the basic format of the USAR.

This figure consists of three columns of data. The first column indicates income statement (P&L) results with $1,000,000 in annual sales with no third-party delivery partner relationship. The second column presents data if a third-party delivery partner is selected, and that partnership generated an additional 10% of the operation's current business. The third column represents the results that would occur if the third-party app produced an additional 50% of the operation's total business.

Foodservice operators making their own careful comparisons related to the financial impact of the costs associated with partnering with third-party delivery apps must make their own reasonable assumptions as they assess the impact of the partnership. The assumptions used in the income statement comparison shown in Figure 7.2 were:

1) The cost of food percentage for on-premises and delivered meals remain unchanged at 30% of sales
2) Total delivery fees and commissions on delivered meals equal 27.5% of sales
3) Total takeout packaging costs for delivered meals equals 5% of delivery sales
4) Increases in management cost to supervise the increased sales levels are small but have been accounted for
5) Variable labor (staff) costs remain unchanged at 30% of total sales

**El Chico Mexican Restaurant**

|  | Column 1 | Column 2 | Column 3 |
|---|---|---|---|
|  | No Third-party | Plus 10% Third-party | Plus 50% Third-party |
| **SALES** |  |  |  |
| Food Sales | $1,000,000 | $1,000,000 | $1,000,000 |
| Third-party Delivery Sales |  | 100,000 | 500,000 |
| **Total Sales** | 1,000,000 | 1,100,000 | 1,500,000 |
| **COST OF SALES** |  |  |  |
| Food @ 30% | 300,000 | 330,000 | 450,000 |
| Delivery fees @ 27.5% |  | 27,500 | 137,500 |
| Takeout Packaging @ 5% |  | 5,000 | 25,000 |
| **Total Cost of Sales** | 300,000 | 362,500 | 612,500 |
| **LABOR** |  |  |  |
| Management | 80,000 | 82,000 | 90,000 |
| Staff @ 30% | 300,000 | 330,000 | 450,000 |
| Employee Benefits @ 17% | 64,600 | 70,040 | 91,800 |
| **Total Labor** | 444,600 | 482,040 | 631,800 |
| **PRIME COST** | 744,600 | 844,540 | 1,244,300 |
| **OTHER CONTROLLABLE EXPENSES** |  |  |  |
| Direct Operating Expenses | 40,000 | 40,000 | 40,000 |
| Marketing | 50,000 | 50,000 | 50,000 |
| Utilities | 10,000 | 10,100 | 10,500 |
| Administrative & General Expenses | 20,000 | 20,000 | 20,000 |
| Repairs & Maintenance | 5,000 | 5,000 | 5,000 |
| **Total Other Controllable Expenses** | 125,000 | 125,100 | 125,500 |
| **CONTROLLABLE INCOME** | 130,400 | 130,360 | 130,200 |
| **NON-CONTROLLABLE EXPENSES** |  |  |  |
| Occupancy Costs | 10,000 | 10,000 | 10,000 |
| Equipment Leases | 0 | 0 | 0 |
| Depreciation & Amortization | 5,000 | 5,000 | 5,000 |
| **Total Non-Controllable Expenses** | 15,000 | 15,000 | 15,000 |
| **RESTAURANT OPERATING INCOME** | 115,400 | 115,360 | 115,200 |
| Interest Expense | 12,000 | 12,000 | 12,000 |
| **INCOME BEFORE INCOME TAXES** | 103,400 | 103,360 | 103,200 |
| Income Taxes @ 20% | 20,680 | 20,672 | 20,640 |
| **NET INCOME** | $ 82,720 | $ 82,688 | $ 82,560 |

Income statement prepared using current (Eighth Edition) Uniform System of Accounts for Restaurants (USAR).

**Figure 7.2** Sales Comparison Income Statement

6) Employee benefits remain unchanged at 17% of total labor costs (cost of management + staff)
7) Other Controllable Costs remain unchanged regardless of volume except utility costs that are small but have been accounted for
8) Non-Controllable Costs remain unchanged
9) Interest Expense remains unchanged
10) Income tax rate of 20% remains unchanged

Given the 10 assumptions cited above, a careful review indicates that an increase in delivery sales of $100,000 (Column 2) or even $500,000 (Column 3) will have little impact on this operation's profits. Using the assumptions indicated, it is not too surprising given that, with a 27.5% delivery fee and a 5% packaging fee, the additional sales generated in this specific example do not make a significant contribution to net income.

It is important to note that, to make a positive impact on its income, the foodservice operator in this example must negotiate a reduction in commission and/or delivery fees, reduce their packaging costs, and/or find savings in other areas (such as reducing its marketing expense).

---

**Find Out More**

It is important for foodservice operators to recognize that those companies offering third-party delivery of the operators' menu items do so with "employee drivers" who are not actually employees. Rather, the third-party delivery drivers are independent contractors. An independent contractor is a person that performs services for another person or entity under a contract between them, with the terms of the relationship spelled out such as duties, pay, and the amount and type of work to be done.

Given the independent contractor relationship, a third-party delivery company cannot *require* one of its contracted drivers to perform a service (i.e., make a delivery). Each independently contracted driver can elect to accept, or not accept, any order the third-party delivery company wants them to deliver.

Increasingly, those who drive for third-party delivery companies are selective in the orders they wish to deliver. The reasons why they are selective are economic. For example, if a requested lunch delivery from a QSR has a price tag of $6.00, a delivery driver will recognize that the gratuity likely to be paid on such an order will be very small. As a result, a driver may simply elect to refuse to deliver that lunch order, while waiting for a delivery request for a larger order.

Similarly, if a delivery order is small, but the distance required to deliver the order is great, an independent contractor can quickly calculate the cost of fuel,

*(Continued)*

vehicle wear and tear, and the time required to make the delivery, and elect to decline delivery of the order.

Increasingly, there are widely reported incidences of multiple smaller orders remaining in a foodservice operation because no driver is willing to deliver them. Foodservice operators should continuously monitor this situation as it is likely to continue to increase in importance in the future. To do that, periodically enter "current restaurant delivery driver employment issues" in your favorite search engine and review the results.

## Managing Service Recovery and Recovery Costs

Even the best managed foodservice operations experience shortfalls in customer service that require appropriate **service recovery** efforts.

The consideration of effective service recovery procedures is best undertaken, and these procedures are important for several reasons:

**Key Term**

**Service recovery:** The actions taken to correct the results of a poor customer service experience (moment of truth) and to regain customer loyalty.

1) *Maintenance of good reputation:* By implementing effective service recovery strategies, foodservice operators are taking steps to ensure their guests are satisfied even when problems occur.
2) *Improved guest retention:* Guests who are satisfied with an operation's service levels are likely to return, even if they have experienced a service-related problem. Note: This is only true if the problem is addressed in a manner that is satisfactory to the guest.
3) *Reduced service recovery costs:* If front-of-house service staff know in advance what to do under a variety of different service recovery strategies, appropriate actions can be determined before they occur. Note: This helps ensure that costs associated with service recovery are known in advance and are appropriately incurred.

Specific service recovery steps in a foodservice operation can range from a sincere apology by a server or manager to monetary compensation to guests including, for example, the **"comping"** of all, or part, of a guest's bill.

Recovery steps related to food quality issues typically involve the replacement, or comping, of a menu item that was, for example, overcooked, undercooked, or served at an incorrect temperature.

Service recovery steps related to service quality issues can best be viewed where they take place.

**Key Term**

**Comp:** The foodservice industry term used to indicate food or beverage items served free of charge to guests or provided to them at prices lower than the regular menu price.

"Comp" is short for "complimentary."

In most cases, foodservice operators must consider the service recovery steps and resulting costs that are undertaken when guests are:

✓ On-premises
✓ Off-premises

## On-Premises Service Recovery

On-premises service recovery means taking care of a service-related shortcoming while the affected guest is on-site. Whenever possible, it is best to attempt to resolve a guest's problem immediately upon becoming aware of it. In some cases, a guest will raise an issue with a service staff member and, in other cases, the guests may request to speak with the manager. In either situation, a prompt response to the guest's concern is essential.

The best method of dealing with a service-related failure varies based on the specific operation in which it occurs and the cause of the service failure. In general, foodservice operators should train staff to select and use one or more of the following service recovery strategies (listed in terms of incremental cost).

1) **Sincere apology:** In some cases, a guest's complaint can be corrected merely by a sincere apology. For example, if a guest requests iced tea to be served with lemon, and the beverage when served does not include a lemon slice, the guest's complaint can normally be addressed simply by apologizing for the service error and assuring the guest that the proper item will be promptly delivered. Apologies cost an operator nothing and, in many cases, serve to properly demonstrate concern for a guest's service-related issue.

2) **Complimentary menu item:** In some cases, an operation can best address a guest's complaint by giving servers the power to "comp" a specific menu item. For example, a server may offer a complimentary appetizer to a table of guests with a confirmed reservation who waited a long time past their reservation time to be seated.

3) **Guest check reduction:** If a menu item has been served, and it did not meet the quality standards the guest anticipated, service recovery can involve the elimination of the menu item from the guest's bill. This can be done whether the guest has consumed the menu item or not and is often a cost-effective way to resolve a service or product deficiency.

4) **Guest check comp:** In the case of a particularly egregious service failure, an operator may resolve the issue by comping a guest's entire check. This service recovery strategy should only be undertaken with the consent of management.

5) **Credit for a future visit:** In some cases, it may be the best course of action to offer a guest a credit to be utilized on a *future* visit. When this approach is taken, a dollar amount should be established. For example, a guest might be given a gift certificate or gift card for $25 off their next meal. Again, this service recovery strategy should only be undertaken with the consent of management.

Service recovery efforts implemented by a foodservice operator should recognize that its goal is to ensure that a guest who has experienced a negative moment of truth leaves the operation satisfied. When they do, they are more likely to return and are less likely to share their negative service experience online.

### Off-Premises Service Recovery

Off-premises service recovery might occur after a guest has left the foodservice operation. In most cases, off-premises service recovery efforts will be taken in response to a negative review of an operation posted online. This demonstrates that problems solved immediately lead to higher online scores because guests with corrected negative experiences are less likely to write about their problem if it was promptly resolved.

Negative reviews typically result in a reduction in the average "score" assigned to a foodservice operation. The average score achieved on a UGC review site is important to those guests deciding whether to visit a specific foodservice operation. Online market research indicates that 70% or more of site visitors reviewing online ratings only continue to read information about an operation if its ratings are within the top two scores (3 and 4 for a four-point ratings system, and 4 and 5 for a five-point rating system). Therefore, scores achieved by an operation directly affect the amount of business they will do.

In fact, a comprehensive study by Professor Michael Luca at the Harvard Business School found that a 0.5-star increase in average Yelp rating generates a 9% revenue increase in revenue for an independent restaurant.[2]

Operators should regularly (several times weekly, if possible) respond to negative (and positive!) online reviews. Creating a dialogue between the operator and customers is perceived positively by those reading online reviews. In addition, responses to reviewers indicates to UGC website visitors that the business considers and values the quality of its customers' experiences as the operation is managed.

The responses to a positive guest review need not be long and detailed. A simple "thank you" and an expression of pleasure that the guests received a positive experience typically suffice.

When they encounter negative reviews posted on UGC review sites, operators should respond to them promptly and professionally for several reasons:

1) *Showing the business cares:* When customers complain, they typically want to know that someone hears them. A response (of almost any kind!) demonstrates that management does care about its guests' challenges and is interested in addressing them.

---

2 https://dl.acm.org/doi/pdf/10.1145/3432953, retrieved August, 30, 2022.

2) *Minimizing the damage:* Once a legitimate negative review has been posted online, it cannot, in most cases, be taken down. However, when readers or viewers of a negative review also immediately encounter a sincere and timely professional management response, the reputational damage done by the negative review is lessened.

3) *Demonstrating professionalism:* A rapid response to negative reviews shows that a business is concerned about its customers, is responsive to them, and will react quickly and professionally when issues arise. These are important traits all consumers look for when they buy products or services.

Foodservice operators do not have the luxury of ignoring negative reviews in the hopes they will go away because, in nearly all cases, they will not. Ignoring negative reviews increases their impact because an operator's failure to reply is not a response to one person. Instead, it is a message to all others who will read or see the review including potential new guests of the operation.

A lack of response by a business tends to confirm the general position of the reviewer (i.e., this business does not care about its customers!). Figure 7.3 lists guidelines for how foodservice operators can positively respond to negative online postings about their business.

When responding to negative (or positive!) reviews, operators must always remember that their responses should be well-written and grammatically correct. The cost of replying to a negative review consists primarily of the responder's time and, in nearly all cases, this is time well spent!

---

1) Thank the reviewer for taking their time to give feedback. (It is always best to start a response in a positive way.)
2) Apologize if the guest is correct about what happened.
3) Avoid arguing if the guest is incorrect about what happened. Instead, apologize if there was a misunderstanding.
4) Provide a very brief but clear and direct explanation of what caused the problem.
5) Assure the site's readers that specific actions have been taken to avoid a repeat of the problem.
6) Offer a direct line of communication between the business and the negative reviewer (via e-mail, text, or phone) to receive a personalized apology.
7) Conclude the response by directly quoting any possible part of the reviewer's comments that were positive (e.g., if the reviewer says the server was "friendly" but the soup was "cold," and mention that it was good to hear the reviewer found the sever to be "friendly.")
8) End the posting by *again* thanking the reviewer for feedback and helping to make the operation even better because of it.
9) Invite the guest to give the operation another chance to provide excellent service and products.

---

**Figure 7.3**  Guidelines for Responding to Negative Online Reviews

The control of service-related costs along with the control of product costs addressed in the previous chapter are important keys to foodservice operation profitability. An additional cost that is of increasing importance to all foodservice operators is that of labor. The control of an operation's labor-related costs is so important it will be the sole topic of the next chapter.

---

**What Would You Do? 7.2**

"I can't believe this guest!" exclaimed Malachi, as he looked over the shoulder of Rita, the owner of Smiley's Pizza, a ghost kitchen that serves takeout pizza and related Italian-style appetizers and desserts. Malachi, the operation's kitchen manager and Rita were looking at Rita's smartphone and reading a guest's online review of their operation. The review had just recently been posted on a popular social media site.

"This guest says that their pizza was fine, but the Fried Mozzarella Sticks they ordered with it were cold and soggy when they were delivered," said Rita.

"I was working in the kitchen when the guest placed the order. We were busy, and I can't remember the exact order. However, I do know we didn't have any problems getting all orders out on time," said Malachi.

"Then I guess we need to look at a potential delivery issue," said Rita. "If our third-party delivery partner's driver picked up more than one order from us that night and, if this guest's order was delivered last, that could explain the problem. Regardless of whose fault it was, the bottom line is that this guest only gave us a 1-star rating. Ouch!"

**Assume you were Rita. Do you think the guest really cares whose fault it was that their mozzarella sticks were cold and soggy? Do you think this guest's complaint was primarily due to a production failure, a service failure, or a takeout menu design failure? What should you do now?**

## Key Terms

| | | |
|---|---|---|
| Third-party delivery | Back-of-house staff | Content management |
| Moment of truth | Ambience |   system (CMS) |
| User-generated content | Dining bubble | Service recovery |
|   (UGC) site | Digital menu board | Comp |
| Ticket times | | |

## Operator's 10-Point Tactics for Success Checklist

Evaluate your need for, and the current status of, each of the following operational tactics. For those tactics you think are important, but not yet in place, develop an action plan for its implementation including who will be responsible for the tactic's completion and the target date by which it should be completed.

| Tactic | Don't Agree (Not Done) | Agree (Done) | Agree (Not Done) | If Not Done | |
|---|---|---|---|---|---|
| | | | | Who Is Responsible? | Target Completion Date |
| 1) Operator understands how the characteristics of intangibility, inseparability, consistency, and limited capacity impact the delivery of high-quality service. | ____ | ____ | ____ | | |
| 2) Operator has carefully considered the importance of managing moments of truth when training staff to provide excellence in guest service. | ____ | ____ | ____ | | |
| 3) Operator recognizes the need to consider speed, accuracy, and professionalism when designing high-quality guest service programs. | ____ | ____ | ____ | | |
| 4) Operator has considered the specific service-related cost control challenges of providing on-premises indoor and outdoor dining. | ____ | ____ | ____ | | |
| 5) Operator has considered the specific service-related cost control challenges of providing drive-thru delivery and guest pick-up of ordered menu items. | ____ | ____ | ____ | | |

*(Continued)*

| Tactic | Don't Agree (Not Done) | Agree (Done) | Agree (Not Done) | If Not Done | |
|---|---|---|---|---|---|
| | | | | Who Is Responsible? | Target Completion Date |
| 6) Operator has considered the specific service-related cost control challenges of providing operator-direct and/or third-party delivery of ordered menu items. | ____ | ____ | ____ | | |
| 7) Operator understands the importance of planning and implementing effective service recovery programs. | ____ | ____ | ____ | | |
| 8) Operator has considered and communicated to service staff the progressive approaches to be taken to address service shortcomings when affected guests are still on-premises. | ____ | ____ | ____ | | |
| 9) Operator recognizes the importance of regularly monitoring online reviews that have been posted by guests who are no longer on-premises. | ____ | ____ | ____ | | |
| 10) Operator recognizes how to professionally respond to negative reviews of their business that have been posted on popular UGC review sites. | ____ | ____ | ____ | | |

# 8

# Labor Cost Control

**What You Will Learn**

1) The Importance of Controlling Labor Costs
2) How to Control Total Labor Costs
3) How to Evaluate Labor Productivity

## Operator's Brief

In this chapter, you will learn about controlling labor costs, an expense that, for most foodservice operations, is second in importance only to that of product costs. In some operations, labor costs actually exceed the amount spent on food and beverage products.

The expenses of management, staff, and benefits comprise an operation's total labor costs. In addition to the cash amounts paid for these items, however, there are numerous non-cash factors that can affect an operation's total cost of labor. Examples include employee selection, training, supervision, scheduling, breaks, the menu, available equipment and tools, and desired levels of service. Each of these non-cash factors is examined in this chapter.

Labor costs can be classified several ways including management and staff, fixed and variable payroll, and controllable and non-controllable labor expenses. Special aspects of labor cost control in the foodservice industry can include addressing issues of regular and overtime pay, child labor laws, and tip accounting. These aspects of labor costs are also addressed in this chapter.

To determine how much labor expense is needed to operate your business, productivity standards must be established. A variety of productivity measures include sales per labor hour, labor dollars per guest served, and the labor cost percentage. Each of these productivity measures are addressed in detail in the chapter including an examination of how they are calculated and their strengths and weaknesses. Finally, in this chapter, you will learn how to optimize your total cost of labor and how to use employee empowerment as a powerful labor cost-reduction tool.

# The Importance of Labor Cost Controls

In most foodservice operations, labor is second only to product costs as the largest incurred expense and in some operations labor costs actually exceed food and beverage product costs. In years past, labor was relatively inexpensive. Today, however, in an increasingly costly labor market, foodservice operators must learn the scheduling and supervisory skills needed to optimize the cost of labor and to maximize the effectiveness of their staff. They must also apply cost control skills to evaluate their efforts.

In some sectors of the foodservice industry, a reputation for long hours, poor pay, and undesirable working conditions has caused some high-quality employees to look elsewhere for more satisfactory careers. It does not have to be that way. When labor costs are adequately controlled, foodservice operators will have the funds necessary to create desirable working conditions and pay wages attractive to the best employees. In every business, including foodservice, better employees mean better guest service and, ultimately, better profits.

### Total Labor Costs

Prime costs (see Chapter 1) in a foodservice operation are the sum of its product (food and beverage) costs and its total labor costs. Therefore, when total labor expense is properly controlled or reduced, prime costs are also reduced.

When utilizing the Uniform System of Accounts for Restaurants (USAR), total labor cost included in an operation's prime cost calculations are the expenses for the operation's management, staff, and employee benefits, as shown in Figure 8.1.

**Figure 8.1** Components of Total Labor Costs (per USAR)

Typically, any employee who receives a salary is considered management. A **salaried employee** receives the same income per week or month regardless of the number of hours worked.

If a salaried employee is paid $1,000 per week when he or she works a complete week, that $1,000 is included in management expense. Salaried employees are more accurately described as **exempt employees** because their duties, responsibilities, and levels of decisions make them "exempt" from the overtime provisions of the U.S. federal government's Fair Labor Standards Act (FLSA).

Exempt employees do not receive overtime for hours worked in excess of more than 40 per week, and they are expected by most foodservice operators to work the number of hours needed to adequately perform their jobs.

Staff costs in a foodservice operation generally refer to the gross pay received by employees in exchange for their work. For example, if an employee earns $18.00 per hour and works 40 hours per week, the gross paycheck (the employee's paycheck before any mandatory or voluntary deductions) would be $720 ($18.00 per hour × 40 hours = $720). This gross amount is recorded as a staff cost.

Employee benefit costs are incurred in every foodservice operation. Mandatory benefits such as required contributions to Social Security are included as an employee benefit cost. Another example is voluntary contributions such as an employer's contribution to an individual employee's 401K retirement account.

The actual total amount of employment taxes and benefits paid by a specific operation can vary greatly. Expenses such as payroll taxes and contributions to workers' unemployment and workers' compensation programs are mandatory for all employers. Other benefit payments such as those for employee insurance and retirement programs are voluntary and vary based on the benefits a foodservice operation chooses to offer its employees. As employment taxes and benefit costs increase, an operation's total labor cost increases, even if its management and staff costs remain constant.

**Key Term**

**Salaried employee:** An employee who regularly receives a predetermined amount of compensation each pay period on a weekly or less frequent basis. The predetermined amount paid is not reduced because of variations in the quality or quantity of the employee's work.

**Key Term**

**Exempt employee:** Employees exempted from the provisions of the Fair Labor Standards Act. These workers typically are paid a salary above a certain level and work in an administrative or professional position. The U.S. Department of Labor (DOL) regularly publishes a duties test that can help employers determine who meets this exemption.

---

**Find Out More**

The difference between exempt and non-exempt employees is clear: Exempt employees are exempt from overtime pay. However, the definition of who qualifies as an exempt employee varies by state.

Employers must accurately classify employees as exempt or non-exempt. Misclassification can result in heavy fines and, without a firm grasp on the distinctions, an employer cannot accurately forecast future payroll costs.

Even experienced foodservice operators can sometimes err in their understanding about the finer details of what makes an employee exempt or non-exempt. To gain a better understanding of who does and does not qualify as an exempt worker in a specific state, enter "exempt worker requirements in (state name)" in your favorite search engine and review the results.

---

### Factors Affecting Total Labor Costs

For some foodservice operators, analysis of total labor costs only involves an assessment of the dollar amount in salaries and staff wages they pay, plus the cost of any benefits they provide.

Experienced foodservice operators, however, recognize that their total labor costs are directly affected by other important non-cash factors such as:

1) Employee selection
2) Training
3) Supervision
4) Scheduling
5) Breaks
6) Menu
7) Equipment/tools
8) Service level provided

#### Employee Selection

Choosing the right employee for a vacant position in a foodservice operation is vitally important in developing a highly productive workforce. Good foodservice operators know that proper employee selection procedures go a long way toward establishing the kind of workforce that is both efficient and cost-effective. This involves matching the right employee with the right job, and the process begins with the development of the **job description.**

**Key Term**

**Job description:** A statement that outlines the specifics of a particular job or position within a foodservice operation. It provides details about the responsibilities and conditions of the job.

A written job description should be maintained for every position in every foodservice operation. From the job description, an appropriate **job specification** can be prepared.

Job descriptions and job specifications are important because they enable foodservice operators to hire only employees who are qualified to

do a job and do it well. Qualified workers can complete their tasks in a more cost-effective manner than will those who are not qualified.

### Training

Perhaps no area under an operator's direct control holds greater promise for increased employee productivity than effective training. In too many cases, however, training in the hospitality industry is poor or almost nonexistent. Highly productive employees are usually well-trained employees, and frequently employees with low productivity have been poorly trained. Every position in a foodservice operation should have a specific, well-developed, and ongoing training program.

Effective training improves job satisfaction and instills in employees a sense of well-being and accomplishment. It will also reduce confusion, product waste, poor service, and loss of guests. In addition, supervisors find that a well-trained workforce is easier to manage than one in which employees are poorly trained.

An operator's training programs need not be elaborate, but they must be consistent and continual. Foodservice employees can be trained in many areas. Skills training allows production employees to understand an operation's menu items and how they are best prepared. Service-related training may be undertaken, for example, to teach new employees how arriving guests should be greeted and properly seated, or how drive-thru orders should be entered in an operation's point-of-sale (POS) system.

It is important to recognize that training must be ongoing to be effective. Employees who are well trained in an operation's policies and procedures should be constantly reminded and updated if their skill and knowledge levels are to remain high. Performance levels can also decline because of a change in the operational systems or changes in equipment used. When these changes occur, employees must be retrained. Effective training costs a small amount of time in the short run but can pay off extremely well in dollar savings in the long term.

### Supervision

All employees require proper supervision, but this is not to say that all employees want to be told what to do. Proper supervision means assisting employees in improving productivity. In this sense, the supervisor is a coach and facilitator who provides employee assistance. Supervising should be a matter of assisting employees to do their best, not just identifying their shortcomings. It is said that employees think one of two things when they see their boss approaching:

1) Here comes help!

or

2) Here comes trouble!

For supervisors whose employees feel that the boss is an asset to their daily routine, productivity gains can be remarkable. Supervisors who only see their positions as one of exercising power, or who see themselves as taskmasters, rarely maintain the quality of workforce needed to compete in today's competitive marketplace.

It is important to remember that it is the employee, not management, who directly services guests. When supervision is geared toward helping employees, then its guests and the entire operation benefits. When employees know that management is committed to providing high-quality menu items and customer service and that it will assist employees in delivering that level of quality, worker productivity will be optimized, and labor costs will be reduced.

### Scheduling

Even with highly productive employees, poor employee scheduling by management can result in low productivity ratios. Consider the example in Figure 8.2, where an operator has determined two alternative schedules for pot washers for an operation that is open for three meals per day.

In Schedule A, four employees are scheduled for 32 hours at a rate of $15.00 per hour. Pot washer payroll in this case is $480 per day (32 hours/day × $15.00/hour = $480 per day). Each shift (breakfast, lunch, and dinner) has two employees scheduled.

In Schedule B, three employees are scheduled for 24 hours. At the same rate of $15.00 per hour, payroll is $360 per day (24 hours per day × $15.00 per hour = $360 per day). Staff costs in this case are reduced by $120 ($480 – $360 = $120), and further savings will be realized because of reduced employment taxes, benefits, employee meal costs, and other labor-related expenses.

Schedule A assumes that the amount of work to be done is identical at all times of the day. Schedule B covers both the lunch and the dinner shifts with two employees, and it also assumes that one pot washer is sufficient in the early-morning period as well as very late in the day.

Scheduling efficiency during the day can often be improved with a **split-shift**, a technique used to match individual employee work shifts with

**Key Term**

**Split-shift:** A working schedule comprising two or more separate periods of duty in the same day.

**Schedule A**

| | 7:30 to 8:30 | 8:30 to 9:30 | 9:30 to 10:30 | 10:30 to 11:30 | 11:30 to 12:30 | 12:30 to 1:30 | 1:30 to 2:30 | 2:30 to 3:30 | 3:30 to 4:30 | 4:30 to 5:30 | 5:30 to 6:30 | 6:30 to 7:30 | 7:30 to 8:30 | 8:30 to 9:30 | 9:30 to 10:30 | 10:30 to 11:30 |
|---|---|---|---|---|---|---|---|---|---|---|---|---|---|---|---|---|
| Employee 1 | ▓ | ▓ | ▓ | ▓ | ▓ | ▓ | ▓ | ▓ | | | | | | | | |
| Employee 2 | ▓ | ▓ | ▓ | ▓ | ▓ | ▓ | ▓ | ▓ | | | | | | | | |
| Employee 3 | | | | | | | | | ▓ | ▓ | ▓ | ▓ | ▓ | ▓ | ▓ | ▓ |
| Employee 4 | | | | | | | | | ▓ | ▓ | ▓ | ▓ | ▓ | ▓ | ▓ | ▓ |

Total Hours = 32

**Schedule B**

| | 7:30 to 8:30 | 8:30 to 9:30 | 9:30 to 10:30 | 10:30 to 11:30 | 11:30 to 12:30 | 12:30 to 1:30 | 1:30 to 2:30 | 2:30 to 3:30 | 3:30 to 4:30 | 4:30 to 5:30 | 5:30 to 6:30 | 6:30 to 7:30 | 7:30 to 8:30 | 8:30 to 9:30 | 9:30 to 10:30 | 10:30 to 11:30 |
|---|---|---|---|---|---|---|---|---|---|---|---|---|---|---|---|---|
| Employee 1 | ▓ | ▓ | ▓ | ▓ | ▓ | ▓ | ▓ | ▓ | | | | | | | | |
| Employee 2 | | | | ▓ | ▓ | ▓ | ▓ | ▓ | ▓ | ▓ | ▓ | | | | | |
| Employee 3 | | | | | | | | | ▓ | ▓ | ▓ | ▓ | ▓ | ▓ | ▓ | ▓ |

Total Hours = 24

**Figure 8.2** Two Alternative Schedules

peaks and valleys of customer demand. When utilizing split-shifts, an operator would, for example, require an employee to work a busy lunch period, be off in the afternoon, and then return to work for the busy dinner period.

Increasingly, foodservice operators can utilize cloud-based scheduling systems designed to allow them to place the right number of workers in the right shifts on the right days and immediately communicate that information to employees with access to the system. Most of these systems also include features that, with management's pre-approval, allow employees to pick up, drop, or swap shifts with other workers. These adaptions give staff more flexibility about when and how much they will work. This type of work flexibility is highly valued by employees and can reduce employee turnover caused by inconvenient worker scheduling. As a result, foodservice operators can also reduce their costs related to staff recruiting and new employee training.

---

**Technology at Work**

In many foodservice operations, the cost of labor is equal to, or even exceeds, the operation's cost of sales. In a foodservice operation, the need for hourly paid employees can vary tremendously at different times of the day and on different days of the week. Therefore, developing and maintaining employee schedules is a critical management task.

The scheduling of hourly employees can be time consuming and challenging as operators plan the employees who should work and when they should work. This is especially so when the operation employs many part-time workers with varying scheduling needs.

Fortunately, some companies have designed software programs that assist operators in creating and quickly modifying hourly worker schedules. The best of these programs allows employees to access the schedule from their own smart devices. Then, there is no need to return to the operation to learn if staff have been added to or reduced from the work schedule.

Employee shift scheduling software saves time and ensures employees are always scheduled according to the operation's needs and the employees' personal preferences and availability. Basic employee scheduling programs are often offered at no cost, and those with more advanced features are available for lease.

To review the features of free-to-use employee scheduling tools, enter "best free scheduling software for restaurants" in your favorite search engine and review the results.

### Breaks

Most employees cannot work at top speed for eight consecutive hours, and they have both a physical and mental need for work breaks. Scheduled short breaks allow them to pause, collect their thoughts, converse with their fellow employees, and prepare for the next work session. Employees given these short breaks will likely produce more than those who are not given any breaks.

Federal law does not mandate that all employees be given breaks, but some states do. As a result, foodservice operators often must determine both the best frequency and length of designated breaks. In some cases, and especially regarding the employment of students and minors, both federal and state laws may mandate special workplace break requirements. Professional operators must be familiar with details about these laws if they apply to their businesses.

### Menu

A major factor in employee productivity is an operation's actual menu. The menu items to be served often have a significant effect on employees' ability to produce the items quickly and efficiently.

In most cases, the greater the number of menu items a kitchen must produce, the less efficient that kitchen will be. Of course, if management does not provide guests with enough choices, loss of sales may result. Clearly, neither too many nor too few menu choices should be offered. The question for operators most often is, "How many selections are too many?" The answer depends on the operation, its employees' skill levels, and the variety of menu items operators believe is necessary to properly attract and serve their guests.

It is extremely important that the menu items selected by management are those items that can be prepared efficiently and serviced well. If done, worker productivity rates will be high, and so will guest satisfaction.

### Equipment/Tools

Foodservice productivity ratios have not increased as much in recent years as have those of other businesses. Much of this is because foodservice is a labor-intensive rather than machine-intensive industry.

In some cases, equipment improvements have made kitchen work easier. Slicers, choppers, and mixers have replaced human labor with mechanical labor. However, in most cases robotics and automation do not significantly contribute to the foodservice industry.

Nonetheless, it is critical for operators to understand the importance of a properly equipped workplace and how it improves productivity. This can be as simple as understanding that a sharp knife cuts more safely, quickly, and better than a dull one or as complex as deciding which Internet system is best to provide data and communication links to the 1,000 stores in a quick-service restaurant chain. In all cases, it is an important part of every operator's job to provide employees with the tools needed to do their jobs quickly and effectively.

### Service Level Provided

The average quick-service restaurant (QSR) employee normally serves more guests in an hour than the fastest server at an exclusive fine dining restaurant. The reason: QSR guests desire speed, not extended levels of rendered services. In contrast, fine dining guests expect more elegant and personal service that is delivered at a much higher level, and this increases the number of necessary employees.

After operators fully recognize the non-cash payment factors that impact their total labor costs, they must next understand the different types of payment-related labor expenses they will incur. Operators must also evaluate the productivity of their workers to determine if their labor dollars were well spent.

## Managing and Controlling Total Labor Costs

The major components of a foodservice operation's total labor cost can be viewed in several ways. Payroll (see Chapter 1) is one major component of every foodservice operation's total labor costs. Some employees are needed simply to open the doors for minimally anticipated business. For example, in a small operation, payroll may include only one supervisor, one server, and one cook. The cost of providing payroll to these three individuals would be the operation's minimum payroll.

Assume, however, that this operation anticipated much greater than minimum business volumes. The increased number of expected guests means that the operation will need more cooks, servers, cashiers, dish room personnel, and perhaps more supervisors to handle the additional workload. These additional positions create a work group that is far larger than the minimum staff, but it is needed to adequately serve the anticipated number of guests. In this scenario, payroll costs will increase.

### Types of Labor Costs

Payroll costs may be viewed several ways, including being either fixed or variable. **Fixed payroll** most often refers to the amount an operation pays in salaries. This amount is typically fixed because it remains unchanged from one pay period to the next unless a salaried employee separates employment from the organization or is given a raise.

**Key Term**

**Fixed payroll:** The amount an operation pays for its salaried workers.

**Variable payroll** consists primarily of those dollars paid to hourly employees (staff) and is an amount that should vary with changes in sales volume. Generally, as sales volume increases, variable payroll expenses increase. In many cases, foodservice operators have little control over their fixed payrolls, but they have nearly 100% control over variable payroll expenses above their minimum staff levels.

Payroll costs may also be viewed as being either controllable or non-controllable. Payroll expenses that are wholly or partially controlled by management are considered **controllable labor expenses**, and labor costs beyond the direct control of management are **non-controllable labor expenses**. Examples of non-controllable labor expenses include employment taxes and some mandatory benefits.

**Key Term**

**Variable payroll:** The amount an operation pays for those workers compensated based on the number of hours worked.

**Key Term**

**Controllable labor expenses:** Those labor expenses under the direct influence of management.

**Key Term**

**Non-controllable labor expenses:** Those labor expenses not typically under the direct influence of management.

## Controlling Total Labor Costs

Properly managing labor costs is important because a variety of laws regulate how and when employees must be paid. For many foodservice operators, particular attention must be paid to two key areas:

✓ Regular and overtime pay
✓ Accounting for tips

### Regular and Overtime Pay

In the United States, the Fair Labor Standards Act (FLSA) establishes minimum wage, overtime pay, recordkeeping, and youth employment standards affecting employees in all businesses employing two or more workers. Covered non-exempt workers are entitled to be paid no less than the Federal minimum per hour. Overtime pay, at a rate not less than one and one-half times the regular rate of pay, is required after 40 hours of work in a workweek. The FLSA sets worker pay requirements in several important areas:

### Minimum wage

The FLSA establishes a minimum hourly rate that must be paid to all employees. If a foodservice operation is located in a state with a mandated minimum wage higher than the federal minimum wage, the state minimum wage must be paid.

## Overtime

Covered non-exempt employees must receive overtime pay for hours worked over 40 per workweek (any fixed and regularly recurring period of 168 hours—seven consecutive 24-hour periods) at a rate not less than one and one-half times the regular rate of pay.

There is no federal limit on the number of hours employees who are 16 years or older may work in any workweek. The FLSA does not require overtime pay for work on weekends, holidays, or regular days of rest unless overtime is worked on such days.

## Hours Worked

Hours worked ordinarily include the time during which an employee is required to be on the employer's premises, on duty, or at a prescribed workplace.

## Recordkeeping

Employers must display an official poster outlining the requirements of the FLSA. Employers must also keep accurate employee time and pay records.

## Child Labor

Child labor provisions set by the FLSA are designed to protect the educational opportunities of minors and prohibit their employment in jobs and under conditions detrimental to their health or well-being.

Every state has its own laws specifically dealing with child labor issues. When federal and state standards are different, the rules that provide the most protection to youth workers apply. U.S. employers must comply with both federal law and applicable state laws.

---

### Find Out More

Many foodservice operators employ young workers. The FLSA generally sets 14 years old as the minimum age for employment, and it limits the number of hours worked by minors under the age of 16.

Special rules also apply to the specific types of tasks these workers can be assigned. Child labor laws also vary by state.

You can better understand the specific requirements of employing young people in a business by making an Internet visit to the department or agency responsible for enforcing youth employment laws in your state. To do so, enter "child labor laws (state name)" in your favorite search engine, and then view the results to learn about any special laws or regulations that apply to minors working in the state.

## Accounting for Tips

Foodservice operations are somewhat unique because they allow employee tips as part of their wage payments when satisfying FLSA worker pay requirements. As a result, a challenge faced by foodservice operators relates to the accounting procedures required to compensate their tipped employees fairly and legally and to keep records of doing so.

In most foodservice operations, a modern POS system is used to record the credit card and cash tips guests intend to give to their servers. Increasingly, however, some foodservice operations use a **tip distribution program** to manage the compensation of their tipped employees.

For example, when a tip distribution program is in place, if a guest gives a server a $45.00 tip on a $200 guest check that included $150.00 of food and $50.00 of alcoholic beverages, the program will assign a portion of the tip to the guest's server and another portion to the bartender(s) who prepared the guest's drinks.

To accurately record tips, tip distribution programs are typically needed when an operation has either a **tip sharing** or **tip pooling** system in place.

The best tip distribution programs are interfaced directly to a restaurant's payroll accounting system. The reason: If an employee "customarily and regularly" makes more than $30 per month in tips, then under current federal law that employee is considered a tipped employee for minimum wage and overtime pay purposes. Many states also have special laws for tipped employees, and some have different standards for qualification as a tipped employee than the federal standard.

### Key Term

**Tip distribution program:** A system of tip payment that allows an operation to distribute a customer's tip from an employee who actually received it to others who also provided service to the customer.

### Key Term

**Tip sharing:** A tip system that takes the tips given to one group of employees and provides a portion of them to another group of employees. Used, for example, when server tips are shared with those who bus the server's tables.

### Key Term

**Tip pooling:** A tip system that takes the tips given to individual employees in a group and shares them equally with all other members of the group. Used, for example, when bartender tips given to an individual bartender are shared equally with all bartenders working the same shift.

**Technology at Work**

Tip distribution software, also referred to as tip pooling or gratuity management software, automates the process of paying tips to tipped workers at the end of their shifts and reduces the need for physical cash payouts.

These tools eliminate hours of manual labor by tracking the number of hours worked and automatically distributing tips to workers' bank accounts. These tools reduce payroll burdens and instances of tip disparity due to an error or theft. They also expedite paying out tips to employees by connecting securely to their bank accounts or debit cards.

Additionally, most tip distribution software allows operators to manage tips across multiple locations and utilize proper reporting features to mitigate risks of non-compliance in tip payments.

To examine some of the offerings of companies that have developed software to assist in tip payment and proper recording, enter "tip distribution software" in your favorite search engine and review the results.

**What Would You Do? 8.1**

Shaheed Kumar is the kitchen manager at the Taron Corporation International Headquarters. The facility he helps manage serves 3,000 employees per day. Shaheed very much needs an additional dishwasher right away. Despite advertising his vacant position for several weeks, Shaheed has had few applicants. He is now interviewing Donny, who is an excellent candidate with five years of experience and who is now washing dishes at the nearby Downriver Rustic Steakhouse.

Shaheed normally starts his new dishwashers at $14.00 per hour. Donny states that he currently makes $16.25 per hour; a rate that is higher than all but one of Shaheed's current dishwashers, many of whom have as much experience as Donny.

As they end the interview, Donny states that, while he would very much like to work at Taron, he simply will not leave his current job if it means he must take a "pay cut."

**Assume you were Shaheed. Would you hire Donny at a pay rate higher than most of your current employees? If so, what would you say to your current dishwashing employees if in the future Donny shared his pay information with them? If not, what would be your plan for filling your vacant dishwasher position?**

## Assessment of Total Labor Costs

To determine how much labor expense is needed to properly operate their businesses, foodservice operators must be able to determine how much work is to be done and how much of it each employee can accomplish. If too few employees are scheduled to work, poor service and reduced sales can result because guests may choose to go elsewhere in search of superior service levels. If too many employees are scheduled, staff wages and employee benefits costs will be too high, and the result will be reduced profits.

**Key Term**

**Productivity (worker):** The amount of work performed by an employee within a fixed time period.

To properly determine the number of workers needed, operators must have a good understanding of the **productivity** of each of their employees.

**Key Term**

**Ratio:** An expression of the relationship between two numbers; computed by dividing one number by the other number.

There are several ways to assess labor productivity. In general, productivity is measured by calculating a productivity **ratio.**

The formula used to calculate a productivity ratio is:

$$\frac{\text{Output}}{\text{Input}} = \text{Productivity Ratio}$$

To illustrate the use of this ratio, assume a foodservice operation employs 4 servers, and it serves 80 guests. Using the productivity ratio formula, the output is guests served, and the input is servers employed:

$$\frac{80 \text{ guests}}{4 \text{ servers}} = 20 \text{ guests per server}$$

This formula states that, for each server employed, 20 guests can be served. The productivity ratio is 20 guests to 1 server (20 to 1) or, stated another way, 1 server per 20 guests (1/20).

There are several ways of defining foodservice output and input, and there are several types of productivity ratios.

Productivity ratios can help an operator determine the answer to the key question, "How much should I spend on labor?" The answer to the question becomes more complex, however, when one recognizes that productivity levels in back-of-house areas (those not open to public access) and front-of-house areas (those open to public access) should, in most cases, be measured differently.

While each foodservice operation is unique, typical measures used to assess *back-of-house* productivity include:

✓ Number of covers (guest orders) completed per labor hour
✓ Number of guest checks (table orders) processed per labor hour
✓ Average guest check completion time (in minutes)
✓ Number of menu items produced per hour worked
✓ Number of improperly cooked items (mistakes) produced per hour worked

Examples of typical measures used to assess the *front-of-house* productivity of servers include:

✓ Number of guests (not tables) served per server hour worked
✓ Number of menu "specials" sold per server hour worked
✓ Number of errors (voided sales) produced per shift worked
✓ Average guest check size (per guest or table served)

Regardless of the productivity measures used, foodservice operators must develop their own methods for managing payroll costs because every foodservice unit is different. Consider, for example, the differences between managing payroll costs incurred by a food truck operator and those required for a large banquet kitchen located in a 1,000-room convention hotel.

Although methods used to manage payroll costs may vary, payroll costs can and must be managed. While there are several ratios operators can use to assess their labor costs (worker productivity), three commonly utilized productivity ratios are:

1) Sales per labor hour
2) Labor dollars per guest served
3) Labor cost percentage

## Sales per Labor Hour

It is said that the most perishable commodity any foodservice operator can buy is the labor hour. When labor is not productively used, it disappears forever. It cannot be "carried over" to the next day as can an unsold head of lettuce or a slice of turkey breast. For this reason, some foodservice operators measure labor productivity in terms of the amount of sales generated for each labor hour used. This productivity measure is referred to as **sales per labor hour**.

**Key Term**

**Sales per labor hour:** The dollar value of sales generated for each labor hour used.

The formula used to calculate sales per labor hour is:

$$\frac{\text{Total Sales Generated}}{\text{Labor Hours Used}} = \text{Sales per Labor Hour}$$

When using this productivity measure, labor hours used is simply the sum of all labor hours paid for by an operation within a specific sales period. To illustrate, consider the operator whose four-week labor usage and the resulting sales per labor hour information is presented in Figure 8.3.

In this example, sales per labor hour ranged from a low of $19.50 in Week 1 to a high of $28.66 in Week 4. Sales per labor hour varies with changes in selling prices, but it will not vary based on changes in prices paid for labor. In other words, increases and decreases in the price paid per hour of labor will not affect this productivity measure. As a result, a foodservice operation paying its employees an average of $15.00 per hour could, using this type of measure for labor productivity, have the same sales per labor hour as a similar unit paying $20.00 for each hour of labor used. Obviously, the operator paying $15.00 per hour has paid far less for an equally productive workforce if the sales per labor hour used are identical in the two units.

Many operators like utilizing the sales per labor hour productivity measure because records on both the numerator (total sales) and the denominator (labor hours used) are readily available. However, depending on the recordkeeping system employed, it may be more difficult to determine total labor hours used than total labor dollars spent. This is especially true when large numbers of managers or supervisors are paid by salary rather than by the hour. Note: It is an operator's choice whether the efforts of both salaried workers and hourly paid staff should be considered when computing an operation's overall sales per labor hour. As long as the operator is consistent with this choice, sales per labor hour from different sales periods can be appropriately compared.

| Week | Total Sales | Labor Hours Used | Sales per Labor Hour |
|---|---|---|---|
| 1 | $18,400 | 943.5 | $19.50 |
| 2 | 21,500 | 1,006.3 | 21.37 |
| 3 | 19,100 | 907.3 | 21.05 |
| 4 | 24,800 | 865.3 | 28.66 |
| Total | $83,800 | 3,722.4 | $22.51 |

**Figure 8.3** Four-Week Sales per Labor Hour

## Labor Dollars per Guest Served

Some foodservice operators measure labor productivity in terms of the labor dollars spent for each guest served. This productivity measure is referred to as **labor dollars per guest served**.

The formula used to calculate labor dollars per guest served is:

### Key Term

**Labor dollars per guest served:** The dollar amount of labor expense spent to serve each of an operation's guests.

$$\frac{\text{Total Cost of Labor}}{\text{Total Number of Guests Served}} = \text{Labor Dollars per Guest Served}$$

To illustrate the use of this ratio, consider the operator whose four-week labor cost and the resulting sales per labor hour information is presented in Figure 8.4.

In this example, the labor dollars expended per guest served for the four-week period would be computed as:

$$\frac{\$29,330}{\$4,190} = \$7.00$$

Note that, in this example, during three weeks (weeks 1–3) the operator provided guests with more than $7.00 of guest-related labor costs per guest served. However, in the fourth week, that amount fell to less than $6.00 per guest. This productivity measure, when averaged, can be useful for operators who find that their labor dollars expended per guest served are lower when their volume is high, and higher when their volume is low.

The utility of labor dollars per guest served is limited in that it varies based on the price paid for labor. Unlike sales per labor hour, however, it is not affected by changes in menu prices.

| Week | Cost of Labor | Guests Served | Labor Dollars per Guest Served |
| --- | --- | --- | --- |
| 1 | $ 7,100 | 920 | $7.72 |
| 2 | 8,050 | 1,075 | 7.49 |
| 3 | 7,258 | 955 | 7.60 |
| 4 | 6,922 | 1,240 | 5.58 |
| Total | $29,330 | 4,190 | $7.00 |

**Figure 8.4** Four-Week Labor Dollars per Guest Served

## Labor Cost Percentage

The most commonly used measure of employee productivity in the foodservice industry is the **labor cost percentage**.

The formula used to calculate a labor cost percentage is:

$$\frac{\text{Total Cost of Labor}}{\text{Total Revenue Generated}} = \text{Labor Cost \%}$$

A labor cost percentage allows an operator to measure the relative cost of labor used to generate a known quantity of sales. It is important to realize, however, that different operators may choose slightly different methods of calculating this popular productivity measure.

Since a foodservice operation's total labor cost consists of management, staff, and employee benefit costs, some operators may calculate their labor cost percentage using only hourly staff wages, or staff wages and salary costs, but not benefit costs. This approach makes sense if an operator can directly control employee pay but not employee benefit costs. It is important to recognize, however, that when operators wish to directly compare their own labor cost percentage to that of other operations, both must have utilized the same formula.

Controlling the labor cost percentage is extremely important in the foodservice industry because it is often used to assess the effectiveness of management. If an operation's labor cost percentage increases beyond what is expected, management will likely be held accountable by the operation's ownership.

Labor cost percentage is a popular measure of productivity, in part, because it is so easy to compute and analyze. To illustrate, consider Roxanne, a foodservice manager in charge of a casual service restaurant in a year-round theme park. The unit is popular and has a $20 per guest check average. Roxanne uses only payroll (staff wages and management salaries) when determining her overall labor cost percentage. The reason: She does not have easy access to the actual amount of taxes and benefits provided to her employees. Roxanne's own supervisor considers these labor-related expenses to be non-controllable and, therefore, beyond Roxanne's immediate influence.

Roxanne has computed her labor cost percentage for each of the last four weeks using her modified labor cost percentage formula. Her supervisor has given Roxanne a goal of 35% labor costs for the four-week period. Roxanne feels that she has done well in meeting that goal. Figure 8.5 shows Roxanne's four-week performance.

| Week | Cost of Labor | Total Sales | Labor Cost % |
|------|---------------|-------------|--------------|
| 1 | $7,100 | $18,400 | 38.6% |
| 2 | 8,050 | 21,500 | 37.4 |
| 3 | 7,258 | 19,100 | 38.0 |
| 4 | 6,922 | 24,800 | 27.9 |
| Total | $29,330 | $83,800 | 35.0 |

**Figure 8.5**   Roxanne's Four Week Labor Cost Percentage Report

Using her labor cost percentage formula and the data in Figure 8.5, Roxanne's Labor Cost % is calculated as:

$$\frac{\text{Cost of Labor}}{\text{Total Sales}} = \text{Labor Cost \%}$$

Or

$$\frac{\$29,330}{\$83,800} = 35\%$$

While Roxanne did achieve a 35% labor cost for the four-week period, Monica, her supervisor, is concerned because she received several negative comments in week 4 regarding poor service levels in Roxanne's unit. Some of these were even posted online, and Monica is concerned about the postings' potential impact on future visitors to the park's foodservice operations. When she analyzes the numbers in Figure 8.5, she sees that Roxanne exceeded her goal of a 35% labor cost in weeks 1 through 3 and then reduced her labor cost to 27.9% in week 4.

Although the monthly overall average of 35% is within budget, Monica knows all is not well in this unit. To achieve her assigned goal, Roxanne elected to reduce her payroll in week 4. However, the negative guest comments suggest that reduced guest service resulted from too few employees on staff to provide the necessary guest attention. As Monica recognized, one disadvantage of using an overall labor cost percentage is that it can hide daily or weekly highs and lows.

In Roxanne's operation, labor costs were too high the first three weeks, and too low in the last week, but she still achieved her overall target of 35%. Roxanne's labor cost of 35% indicates that, for each dollar of sales generated, 35 cents was paid to the employees who assisted in generating those sales. In many cases, a targeted labor cost percentage is viewed as a measure of employee productivity and, to some degree, management's skill in controlling labor costs.

While it is popular, in addition to its tendency to mask productivity highs and lows, the labor cost percentage has some limitations as a measure of productivity.

| Week | Original Cost of Labor | 5% Pay Increase | Cost of Labor (with 5% Pay Increase) | Total Sales | Labor Cost % |
|---|---|---|---|---|---|
| 1 | $7,100 | $355.00 | $7,455.00 | $18,400 | 40.5% |
| 2 | 8,050 | 402.50 | 8,452.50 | 21,500 | 39.3 |
| 3 | 7,258 | 362.90 | 7,620.90 | 19,100 | 39.9 |
| 4 | 6,922 | 346.10 | 7,268.10 | 24,800 | 29.3 |
| Total | 29,330 | 1,466.50 | 30,796.50 | 83,800 | 36.8 |

**Figure 8.6** Roxanne's Four-Week Revised Labor Cost % Report (Includes 5% Pay Increase)

Note, for example, what happens to this measure of productivity if all of Roxanne's employees are given a 5% raise in pay. If this were the case, her labor cost percentages for last month would be calculated as shown in Figure 8.6.

Note that labor now accounts for 36.8% of each sales dollar, but one should realize that Roxanne's workforce did not become less productive simply because they got a 5% increase in pay. Rather, the labor cost percentage changed due to a difference in the price paid for labor. When the price paid for labor increases, labor cost percentage increases and, similarly, when the price paid for labor decreases, the labor cost percentage decreases. Therefore, using the labor cost percentage alone to evaluate workforce productivity can sometimes be misleading.

Another example of the limitations of the labor cost percentage as a measure of labor productivity can be seen when selling prices are increased. Return to the data in Figure 8.5 and assume that Roxanne's unit raised all menu prices by 5% effective at the beginning of the month. Figure 8.7 shows how this increase in her selling prices would affect her labor cost percentage.

| Week | Cost of Labor | Original Sales | 5% Selling Price Increase | Sales (with 5% Selling Price Increase) | Labor Cost % |
|---|---|---|---|---|---|
| 1 | $7,100 | $18,400 | $920 | $19,320 | 36.7% |
| 2 | 8,050 | 21,500 | 1,075 | 22,575 | 35.7 |
| 3 | 7,258 | 19,100 | 955 | 20,055 | 36.2 |
| 4 | 6,922 | 24,800 | 1,240 | 26,040 | 26.6 |
| Total | 29,330 | 83,800 | 4,190 | 87,990 | 33.3 |

**Figure 8.7** Roxanne's Four-Week Revised Labor Cost % Report (Includes 5% Increase in Selling Price)

Note that increases in selling prices (assuming no decline in guest count or changes in guests' buying behavior) will result in *decreases* in the labor cost percentage. Alternatively, lowering selling prices without increasing total revenue by an equal amount will result in *increases* in labor cost percentage.

Although labor cost percentage is easy to compute and widely used, it is difficult to use as a measure of productivity over time. The reason: It depends on labor dollars spent and sales dollars received for its computation. Even in relatively noninflationary times, wages do increase, and menu prices are adjusted. Both activities directly affect the labor cost percentage, but not worker productivity. In addition, institutional foodservice settings, which often have no daily dollar sales figures to report, can find that it is not easy to measure labor productivity using labor cost percentages because operators generally calculate and report guest counts or number of meals served rather than sales dollars earned.

Figure 8.8 summarizes key characteristics of the three measures of labor productivity presented.

Regardless of the productivity measure utilized, if an operator finds labor costs are too high relative to sales produced, problem areas must be identified, and corrective action(s) must be taken. If the overall productivity of employees cannot be improved, other action(s) becomes important.

The approaches operators can take to reduce labor-related costs are different for fixed payroll costs than for variable payroll costs. Figure 8.9 summarizes strategies

| Measurement | Advantages | Disadvantages |
| --- | --- | --- |
| Sales per Labor Hour = $\frac{\text{Total Sales}}{\text{Labor Hours Used}}$ | 1) Fairly easy to compute 2) Does not vary with changes in the price of labor | 1) Ignores price per hour paid for labor 2) Varies with changes in menu selling price |
| Labor Dollars per Guest Served = $\frac{\text{Cost of Labor}}{\text{Guests Served}}$ | 1) Fairly easy to compute 2) Does not vary with changes in menu selling price 3) Can be used by non-revenue-generating units | 1) Ignores average sales per guest and, therefore, total sales 2) Varies with changes in the price of labor |
| Labor Cost% = $\frac{\text{Cost of Labor}}{\text{Total Sales}}$ | 1) Easy to compute 2) Most widely used | 1) Hides highs and lows 2) Varies with changes in the price of labor 3) Varies with changes in menu selling price |

**Figure 8.8** Productivity Measures Summary

| Labor Category | Actions |
|---|---|
| Fixed | 1) Increase sales volume.<br>2) Combine jobs to eliminate fixed positions.<br>3) Reduce wages paid to fixed-payroll employees. |
| Variable | 1) Improve productivity.<br>2) Schedule appropriately to adjust to changes in sales volume.<br>3) Combine jobs to eliminate variable positions.<br>4) Reduce wages paid to variable employees. |

**Figure 8.9**   Reducing Labor-Related Expense

operators can use to reduce labor-related expense percentages in each of these two categories. Note that operators can only decrease variable payroll expense by increasing productivity, improving the scheduling process, eliminating employees, or reducing wages paid.

Another often ignored tactic that can increase employee productivity and reduce labor-related expense is **employee empowerment**. Employee empowerment results from a decision by management to fully involve employees in the decision-making process as far as guests and the employees themselves are concerned.

Many experienced managers remember that it was once customary for management to (a) make all decisions regarding every facet of the operational aspects of its organization and (b) present them to employees as inescapable facts to be accomplished. Instead, an alternative approach occurs when employees are given the "power" to get involved.

**Key Term**

**Empowerment (employee):** An operating philosophy that emphasizes the importance of allowing employees to make independent guest-related decisions and to act on them.

Employees can be empowered to make critical decisions concerning themselves and, most importantly, an operation's guests. As addressed in Chapter 7, many front-of-house employees work closely with guests, and numerous problems are more easily solved when employees are given the power to make it "right" for the guests. Successful operators often find that well-planned and consistently delivered training programs can be helpful. Empowered employees can also yield a loyal and committed workforce that is more productive, is supportive of management, and will "go the extra mile" for guests. Doing so helps reduce labor-related costs, builds repeat sales, and increases profits.

The control of food and beverage product costs, as well as labor costs, are extremely important in a foodservice operator's overall cost control efforts. There are, however, other expenses that significantly affect the profitability of a foodservice operation. It is the recognition and control of these other expenses that are the topic of the next chapter.

**What Would You Do? 8.2**

"I'm just saying that we are already reporting our daily labor cost percentage to the owners, and now they also want to know this every day. It just seems like more paperwork to me," said Leon Sullivan, the assistant manager of the Philadelphia Sliders restaurant.

Leon was talking to Ruth, the restaurant's manager. Ruth had just informed Leon that the restaurant's owners wanted her to submit a daily "Average Drive-through Ticket Time Report" along with the day's revenue and labor cost percentage reports.

Along with on-site dining, the Philadelphia Sliders drive-thru was seeing an increasing amount of business. However, there have been some complaints about slow service. In response, the restaurant's owners requested that Ruth calculate and report an average drive-thru ticket time. Note: Ticket time begins when a guest's order is placed, and it ends when the guest leaves the drive-through window with their ordered items.

The restaurant's POS system was programmed to create a unique guest check for each order, and the system recorded the time at which a guest's order was placed and when the order was completed. One reporting feature of the POS system was that it automatically calculated the average drive-thru ticket time for all drive-thru orders, and this was the information the operation's owners now wanted to review each day.

**Assume you were Ruth. Why do you think this operation's owners are now requesting this new employee productivity report? How much do you believe the information contained in this report will help your labor cost control efforts, and your efforts to optimize guest service in the drive-thru?**

## Key Terms

| | | |
|---|---|---|
| Salaried employee | Controllable labor | Ratio |
| Exempt | expenses | Sales per labor hour |
| employee | Non-controllable labor | Labor dollars per |
| Job description | expenses | guest served |
| Job specification | Tip distribution program | Labor cost percentage |
| Split-shift | Tip sharing | Empowerment |
| Fixed payroll | Tip pooling | (employee) |
| Variable payroll | Productivity (worker) | |

## Operator's 10-Point Tactics for Success Checklist

Evaluate your need for, and the current status of, each of the following operational tactics. For those tactics you think are important, but not yet in place, develop an action plan for its implementation including who will be responsible for the tactic's completion and the target date by which it should be completed.

| Tactic | Don't Agree (Not Done) | Agree (Done) | Agree (Not Done) | Who Is Responsible? | Target Completion Date |
|---|---|---|---|---|---|
| 1) Operator understands the relationship between profits and controlling labor costs. | —— | —— | —— | | |
| 2) Operator understands the difference between fixed payroll and variable payroll. | —— | —— | —— | | |
| 3) Operator recognizes the basic difference between controllable and non-controllable labor expenses. | —— | —— | —— | | |
| 4) Operator can identify the requirements of the Fair Labor Standards Act (FLSA) that apply to their own business. | —— | —— | —— | | |
| 5) Operator recognizes the unique requirements affecting wage payments to tipped employees. | —— | —— | —— | | |
| 6) Operator knows how to calculate and assess the "Sales per Labor Dollar" measure of productivity. | —— | —— | —— | | |
| 7) Operator knows how to calculate and assess the "Labor Dollars per Guest Served" measure of productivity. | —— | —— | —— | | |
| 8) Operator knows how to calculate and assess the "Labor Cost Percentage." | —— | —— | —— | | |

*(Continued)*

| Tactic | Don't Agree (Not Done) | Agree (Done) | Agree (Not Done) | If Not Done | |
|---|---|---|---|---|---|
| | | | | Who Is Responsible? | Target Completion Date |
| 9) Operator understands the various actions that can be taken to reduce fixed and variable labor-related expenses. | ____ | ____ | ____ | | |
| 10) Operator recognizes the impact employee empowerment can have on reducing labor costs and improving profits. | ____ | ____ | ____ | | |

# 9

# Control of Other Operating Expenses

---

**What You Will Learn**

1) The Importance of Controlling Other Expenses
2) How to Classify Other Expenses as Controllable or Non-controllable
3) How to Categorize Other Expenses in Terms of Being Fixed, Variable, Or Mixed
4) How to Compute Other Expense Costs as a Percentage of Total Sales and on a Cost Per-Guest Basis

---

**Operator's Brief**

In addition to the effective control of products, service, and labor costs, all foodservice operators must be concerned about other expenses required to operate their businesses. In this chapter, you will learn the importance of managing other operating expenses.

In some cases, other expenses will be controllable by you, and in other cases these costs are considered non-controllable. In this chapter, you will learn the differences between these two types of costs.

Other expenses can also be classified as either fixed, variable, or mixed. In this chapter, you will learn about the important differences between these types of costs and how their control can help you reach your financial goals.

While the control of all other expenses is important, specific areas of particular importance to foodservice operators are those other expenses associated with food and beverage operations, equipment maintenance and repair, technology-related costs, and occupancy expenses. Each of these key cost areas are addressed in this chapter.

As you control other expense costs, you can employ two important cost control tools. These are the "other expense cost percentage" and the "other expense cost per guest." Both are important tools depending on the specific information you seek. In this chapter, you will learn how to calculate and assess these important other expense control metrics.

---

**CHAPTER OUTLINE**

The Importance of Controlling Other Operating Expenses
Controllable and Non-controllable Other Expenses
Fixed, Variable, and Mixed Other Expenses
Monitoring and Controlling Other Expenses
    Food and Beverage Operations
    Equipment Maintenance and Repair
    Technology-related Costs
    Occupancy Costs
Calculating Other Expense Costs
    Other Expense Cost Percentage
    Other Expense Cost Per-guest

---

## The Importance of Controlling Other Operating Expenses

In nearly all cases, food, beverage, and labor expenses represent the largest cost areas most foodservice operators will encounter. In addition to food, beverage, and labor expenses, however, there are "other" operating expenses that every operator must pay and control.

These other expenses can account for a significant amount of the total cost of operating a foodservice business, and they include a variety of items. Some common examples include advertising, website maintenance, utility bills, equipment repair, liability insurance, property taxes, and mortgage payments (if a building is owned by the operation) or rent/lease payments (if the building is owned by others).

Some of these other expenses are directly under management control. When following the Uniform System of Accounts for Restaurants (USAR), these types of expenses are reported on an operation's income statement as "Other Controllable Expenses" (see Chapter 1). Some other expenses, however, are not directly controllable by management and, as a result, these are reported on the income statement as "Non-controllable Expenses."

Managing a foodservice operation's other expenses is just as important to its success as controlling its food, beverage, and payroll expenses. In most cases, the profit margins in many foodservice operations are small, so the control of all costs is critically important. Even in those situations that are traditionally considered non-profit such as hospitals and educational institutions, all operating costs must be controlled. The reason: Dollars that are wasted in foodservice will not be available for use in other important areas of the facility.

Effective foodservice operators continually look for ways to control all their operating expenses, but sometimes the environment in which a business operates influences some operating costs in positive or in negative ways.

One good example of this relates to energy conservation and waste recycling. Energy costs are one of the other expenses examined in this chapter. In the past, a table service restaurant serving water to each guest upon arrival was simply a **standard operating procedure (SOP)**. However, the rising cost of energy and an increased awareness about the environment of wasted resources have generated a new policy: serve water only upon request instead of to each arriving guest.

**Key Term**

**Standard operating procedure (SOP):** The term used to describe how something should be done under normal business operating conditions.

In most cases, guests have found this change acceptable, and the expense savings related to glass washing, equipment usage, energy, cleaning supplies, and labor costs can be significant. Note: In a similar manner, many operators today are finding that recycling fats and oils, cans, jars, and paper products can be good for the environment and also for the property's bottom lines. Recycling these items reduces an operation's cost of routine trash disposal and, in some communities, the recycled materials themselves have a cash value.

## Controllable and Non-controllable Other Expenses

It is most often helpful for foodservice operators to consider their various other expenses in terms of the costs being either controllable or non-controllable. To see why, consider the case of Sven Hedin, the owner of a neighborhood tavern. Most of Sven's sales revenue comes from the sale of beer, sandwiches, and his special pizza. Sven is free to decide on the monthly amount he will spend on advertising. Advertising expense, then, is under Sven's direct control and is considered a controllable other expense.

Some of his other expenses, however, are not under his direct control. One example is the license needed to operate his facility legally. The state in which Sven operates charges a liquor license fee to all those businesses that serve alcoholic beverages. If his state increases the liquor license fee, Sven is required to pay the additional fee. In this situation, the alcoholic beverage license fee is considered a non-controllable expense; that is, an expense that is outside of Sven's direct control.

As an additional example, assume an operator's business was part of a franchised quick-service chain that sells chicken sandwiches and chicken strips. Each month, the operator is assessed a $1,000 advertising and promotion fee by the chain's regional headquarters. The $1,000 is used to purchase television advertising time for the chain. This $1,000 charge is a non-controllable operating expense because it is set by the franchisor, and it must be paid each month regardless of the revenue level achieved by the operator.

A controllable expense is one in which decisions made by a foodservice operator can have the effect of either increasing or reducing the expense. A non-controllable expense is one that the operator can neither readily increase nor decrease. Operators have some control over controllable expenses, but they usually have little or no control over non-controllable expenses. As a result, operators should focus their attention primarily on controllable (not non-controllable) expenses.

It is also important to recognize that the items categorized as other expenses can constitute almost anything in the foodservice business. If a restaurant is a floating ship, periodically scraping the barnacles off the boat is categorized as other expense! If an operator is serving food to oil field workers in Alaska, heating fuel for the dining rooms and kitchen will be another expense, and probably a very large one! If a company has been selected to cater an outdoor wedding, tent rental and tent erection costs may be significant other expenses.

Each foodservice operation will have its own unique list of required other expenses. The USAR actually lists several hundred other expense categories commonly incurred by foodservice operations. It is not possible, therefore, to list all imaginable expenses that could be incurred by every foodservice operator. The expenses that are incurred, however, should be recorded and reported in a meaningful way. For example, some restaurants incur napkin, straw, paper cup, and plastic lid costs. All these costs could be conveniently reported under a general grouping or cost category such as "Paper supplies and packaging."

Similarly, nearly all bars will incur expense for items such as jiggers, stir sticks, paper coasters, tiny plastic swords (for drink garnishes), and small paper umbrellas. These costs might be combined and reported in a cost category such as "Bar utensils and supplies."

When other expense cost groupings are used, they should make sense to the operator and be specific enough to let the operator know exactly what items are in the category. Although some operators prefer to make up their own other expense cost categories, the cost categories suggested in this book are recommended by the USAR. The individual other expense cost categories generally recommended by the USAR are divided into two primary groups:

1) Other Controllable Expenses
2) Non-controllable Expenses

Within each of these two major other expense groups, recommended groupings for individual cost categories are:

1) Other Controllable Expenses
   - Direct Operating Expenses
   - Music and Entertainment
   - Marketing
   - Utilities
   - General and Administrative Expenses
   - Repairs and Maintenance
2) Non-controllable Expenses
   - Occupancy Costs
   - Equipment Rental
   - Depreciation and Amortization
   - Corporate Overhead (multi-unit restaurants)
   - Management Fees
   - Interest Expense

Within each of the two primary other expense cost subcategories, an operation's individual other expenses can be even further detailed to assist operators in their reporting and decision-making activities. For example, the USAR suggests the following individual costs be reported under the "Other Controllable Expenses" subcategory of "Utilities":

✓ Electricity
✓ Gas
✓ Heating Oil and Other Fuel
✓ Recycling Credits
✓ Trash Removal
✓ Water and Sewage

Similarly, those expenses that are non-controllable may be detailed within their individual cost categories. For example, the USAR suggests the following individual costs be reported under the "Non-controllable Expenses" subcategory of "Occupancy Costs" (see Chapter 1):

✓ Rent—Minimum or Fixed Amount
✓ Rent—Percentage Rent Amount
✓ Rent—Parking Spaces
✓ Ground Rent
✓ Common Area Maintenance
✓ Insurance on Building and Contents
✓ Other Municipal Taxes

✓ Personal Property Taxes
✓ Real Estate Taxes

One major purpose of the USAR is to provide operators and owners with guidance on how to best report the individual other expenses their businesses incur. The reasons why this is important will become very clear when operators analyze their cost control efforts using the income statement (see Chapter 12).

## Fixed, Variable, and Mixed Other Expenses

While it is important to recognize costs as either controllable or non-controllable, it is also important to understand that some of the other expenses to be paid in a foodservice operation will be the same amount each month.

For example, if an operator leases a building to house a restaurant, the lease payment will likely be the same amount each month. In some other instances, the amount of an expense will vary based on the success of the operation. Because each guest served receives at least one napkin, the expenses an operator incurs for paper cocktail napkins used in a cocktail lounge increase as the number of guests served increases and decrease as the number of guests served decreases. Figure 9.1 summarizes the impact on each of these types of cost as sales volume varies.

A convenient way to remember the differences between fixed, variable, and mixed expenses is to consider the paper napkins in a napkin holder placed at the beginning of a cafeteria line. The napkin holder is a fixed expense. One holder is sufficient whether an operator serves 10 guests at lunch or 50 guests during this meal. The napkins themselves, however, are a variable expense. As more guests are served (assuming each guest takes one napkin), the cafeteria will incur a greater paper napkin expense. The cost of the napkin holder and the napkins, if considered together, would be considered a mixed expense.

Effective managers know they should not categorize controllable, non-controllable, fixed, variable, or mixed costs in terms of being either "good" or "bad." Some expenses are, by their nature, related to sales volume, and others are not. It is important to remember that, in most cases, the goal of management is

| Expense | As a Percentage of Sales | Total Dollars Spent |
| --- | --- | --- |
| Fixed Expense | Decreases | Remains the Same |
| Variable Expense | Remains the Same | Increases |
| Mixed Expense | Decreases | Increases |

**Figure 9.1** Fixed, Variable, and Mixed Expense Behaviors as Sales Volume Increases

not to reduce but rather to increase variable expenses in direct relation to increases in sales volume.

Expenses are required to service guests. In the example of the paper napkins, it is clear managers would prefer to use 50 paper napkins at lunch rather than 10 napkins. While the total cost of servicing guests is less than the amount they are charged, increasing the number of guests served will increase variable other expenses, and it will increase profits as well.

## Monitoring and Controlling Other Expenses

Remember that each foodservice operation faces its own unique set of other expenses. For example, a restaurant on a southern Florida beach may well consider the expense of purchasing hurricane insurance, while a similar restaurant in Colorado would not consider this expense. Each foodservice operation is unique.

Effective operators are constantly on the lookout for ways to reduce unnecessary additions to any of their other expense categories. For most operators, however, the effective control of other expenses must include special attention to four key other expense areas:

✓ Food and Beverage Operations
✓ Equipment Maintenance and Repair
✓ Technology-related Costs
✓ Occupancy Costs

### Food and Beverage Operations

In many respects, the other expenses related to food and beverage operations should be treated just like food and beverage expenses. For instance, in the case of cleaning supplies, linens, uniforms, and the like, products should be ordered, inventoried, and carefully issued in the same manner used for food and beverage products.

Recall that, in most cases, reducing the total dollar amount of variable cost expenses is generally not desirable because, in fact, each additional sale creates additional variable expenses. In this case, although total variable expenses may increase, the positive impact of the additional sales on fixed costs will serve to reduce an operation's overall other expense percentage. It is also true that an operation's fixed costs can only be reduced when they are measured as a percentage of total sales. This, in turn, is achieved by increasing an operation's total sales.

| Sales | Fixed Expense | Variable Expense (10%) | Total Other Expense | Other Expense Cost % |
|---|---|---|---|---|
| $ 1,000 | $150 | $ 100 | $ 250 | 25.00% |
| $ 3,000 | $150 | $ 300 | $ 450 | 15.00% |
| $ 9,000 | $150 | $ 900 | $1,050 | 11.67% |
| $10,000 | $150 | $1,000 | $1,150 | 11.50% |
| $15,000 | $150 | $1,500 | $1,650 | 11.00% |

**Figure 9.2**   Igloo's Fixed and Variable Other Expenses

| Activity | Energy Use |
|---|---|
| Cooking | 32% |
| Heating (building) | 19% |
| Cooling (building) | 18% |
| Heating (water) | 13% |
| Refrigeration | 11% |
| Lighting | 6% |
| Administrative | 1% |
| **Total** | **100%** |

**Figure 9.3**   Typical Restaurant's Energy Consumption Pattern

To illustrate, consider a shaved-ice kiosk called Igloo's located in the middle of a small mall's parking lot. Figure 9.2 shows the impact of volume increases on both total other expense and other expense cost percentage. In this example, some of the other expenses related to food and beverage operations are fixed and others are variable. The variable portion of other expense, in this example, equals 10% of gross sales. Fixed expenses equal $150.

Note that the variable expense shown in Figure 9.2 increases from $100 to $1,500 as sales increase from $1,000 to $15,000. However, when that occurs, total other expense percentage drops from 25% of sales to 11% of sales. Therefore, operators desiring to reduce their percentage of other expense costs directly related to food and beverage operations must increase their sales.

Utility services costs are an extremely important category of other expenses related to food and beverage operations. To produce their menu items, serve, and clean up, foodservice operators may use thousands of gallons of water. They may also consume significant amounts of natural gas (generally used for cooking and water heating) and utilize a large number of **kilowatt hours (kwh)** each month.

**Key Term**

**Kilowatt hour (kwh):** A measure of electrical usage.

Like product and labor costs, utility costs can be controlled. This process starts by understanding just where an operation uses its energy. Although the heating and cooling costs incurred by a restaurant in Alaska will be different than those of a restaurant in Arizona, the usage pattern shown in Figure 9.3 is a typical one.

---

**Find Out More**

According to the U.S. Department of Energy, restaurants use about five to seven times more energy per square foot than other commercial buildings. High-volume quick-service restaurants (QSRs) may even use up to 10 times more energy per square foot than other commercial buildings.*

ENERGY STAR® is a joint program of the Environmental Protection Agency (EPA) and the Department of Energy (DOE). Its goal is to help consumers, businesses, and industry save money and protect the environment by adopting energy-efficient products and practices. The ENERGY STAR label identifies top-performing, cost-effective products, homes, and buildings.

The Department of Energy has also produced the *ENERGY STAR Guide for Cafés, Restaurants, and Institutional Kitchens* to help operators identify ways to save energy and water in businesses, boost their bottom lines, and help protect the environment. This resource also contains tips on how to upgrade equipment and highlights best practices that can positively impact a business's daily operations. This guide specifically addresses ENERGY STAR for Commercial Foodservice and other energy-saving options including Lamps and Lighting Fixtures, Heating, Ventilating and Air Conditioning (HVAC), and Water and Waste Management.

To learn more about money-saving energy-related tips, enter "ENERGY STAR Guide for Cafés, Restaurants, and Institutional Kitchens" in your favorite search engine and review the results.

* https://www.energystar.gov/buildings/resources_audience/small_biz/restaurants, retrieved March 30, 2023.

---

A foodservice operation's utility costs can (and should!) be well-managed, and teaching staff about the information in Figure 9.4 is a good way to begin the process.

### Equipment Maintenance and Repair

Every skilled craftsperson knows that keeping work tools clean and in good working order will make the tools last longer and perform better. The same is true for tools and equipment in foodservice facilities. Proper care of mechanical equipment not only prolongs its life but also reduces operating costs. As prices for water, gas, and other energy sources needed to operate facilities continue to rise, effective foodservice operators must implement a facility repair and maintenance program that seeks to discover and treat minor equipment and facility problems before they become major problems.

1) Turn It Off
   - Turn off lights, cooking equipment, and exhaust fans when they are not being used.
   - Activate the standby mode for office equipment, in-house computers, and printers to effectively put these pieces of equipment "to sleep" when not in use.
2) Keep It Closed
   - Keep refrigerator doors closed.
   - Keep outside doors, if any, to the kitchen closed to minimize heat and cooling loss.
3) Turn It Down
   - Set air-conditioning units at 76°F (24.5°C) for cooling.
   - Set heating systems at 68°F (20°C) for heating.
   - Reduce the temperature of water heaters (where appropriate).
   - Adjust heating/cooling temperature settings when the operation is closed for the night.
4) Vent It
   - Use ceiling fans to help recirculate dining room air.
   - Retrofit exhaust hoods with both low- and high-speed fans in the dish room, food preparation, and cooking areas.
5) Change the Bulbs
   - Replace incandescent bulbs with LED or fluorescent bulbs because they use less electricity and last many times longer.
   - Install photocell light sensors (motion detectors) where appropriate (examples: storage areas and the like) to activate lighting only when needed.
6) Watch the Water
   - Run dishwashers only when they are full.
   - Replace/repair leaking faucets immediately.
   - Insulate all hot water pipes.
   - Install "water-saver" spray nozzles in dish areas.
7) Cook Right
   - Stagger preheat times for equipment to minimize surcharges for high energy use.
   - Bake during off-peak periods (contact the local electrical company to learn about off-hour usage).
   - Idle cooking equipment between meal periods at reduced temperatures where appropriate.
8) Seal It
   - Caulk and weather-strip cracks and openings around doors, windows, vents, and utility outlets.
   - Check freezer, refrigerator, and walk-in seals and gaskets for cracks or warping. Replace as needed.
9) Maintain It
   - Change air filters on a regular basis (monthly during peak heating and cooling seasons).
   - Clean grease traps on ventilation equipment.
   - Clean air-conditioner and refrigeration condenser/evaporator coils at least every three months.
   - Oil, lube, clean, and repair equipment as needed to maximize operating efficiency.
10) Get Help
   - Take advantage of any advisory services offered by local utility companies and governmental agencies.
   - Regularly consult with the operation's heating, ventilation, and air-conditioning (HVAC) repair person for tips on minimizing energy and maintenance costs with the business's particular HVAC system. It's like getting a free energy management consultant!

**Figure 9.4** Ten Common Sense Energy Tips for Foodservice Operators

One way to help ensure that costs are as low as possible is to use a competitive-bid process before awarding contracts for required services. For example, if an operator hires a carpet cleaning company to clean dining room carpets monthly, it is a good idea to annually seek competitive bids from new carpet cleaners. This can help to reduce costs by ensuring that the carpet cleaner selected has given a price that is competitive with other service providers in the area. For general maintenance contracts in areas such as the kitchen or for mechanical equipment, elevators, or grounds maintenance, it is a good idea to get bids for these services at least once per year. This is especially true if the dollar value of the services contract is large.

A second important equipment maintenance and repair-related task is the creation and use of a **maintenance log**. An equipment maintenance log is a document that lists all service-related actions that have been taken on a piece of equipment. As such, it serves as a history of the equipment's maintenance and service at the property.

**Key Term**

**Maintenance log:** A document that contains detailed information about equipment and the actions (e.g., oiling, filter replacement, and repair) that has been performed on it.

The use of equipment logs allows foodservice operators to manage their equipment checkup and maintenance tasks, record when the tests were performed, and document each task's purpose.

Foodservice operators can rely on equipment maintenance logs to ensure that their most important pieces of equipment are kept in good working order and that the chance of their operations being affected by unplanned equipment downtime is minimized. Without an appropriate maintenance tracking system, it is likely that maintenance schedules can be missed, and maintenance needs can go unfulfilled.

To help control equipment maintenance and repair costs, all air-conditioning, plumbing, heating, and refrigerating units should be inspected at least yearly. In addition, kitchen equipment such as dishwashers, slicers, and mixers should be inspected at least monthly for preventive maintenance purposes.

Some foodservice managers operate facilities that are large enough to employ their own full-time maintenance staff. In these cases, it is important to ensure these employees have copies of the operating and maintenance manuals of all equipment. These documents can prove invaluable in the reduction of equipment and facility-related operating and repair costs. (Note: Increasingly, this information is found on the Internet, and\or the equipment's local dealer can be contacted to obtain this or similar information.)

---

**Technology at Work**

Maintenance logs need not be complex nor hard to maintain. In most cases, the logs will consist of two key parts.

The first part provides general equipment-related information. This information is used to identify each individual piece of equipment and will commonly include:

1) Name of equipment
2) Model or manufacturer
3) Serial number
4) Location
5) Person responsible for scheduling equipment work
6) Purchase date

The second section lists all the maintenance actions performed on the equipment and will commonly include:

1) The date when the action was performed
2) Description of the maintenance
3) Name and contact information of the person performing the maintenance
4) Cost of the action taken

Fortunately, several companies produce maintenance log templates that can be downloaded for free or at very little cost. To examine some of these, enter "free maintenance log template for foodservice operators" in your favorite search engine and view the results.

---

## Technology-related Costs

Technology-related costs refer to those expenses incurred by a foodservice operation that are directly related to a variety of hardware, software, and apps that are used to improve customer service and reduce operating costs. While friendly smiles, good food, and prompt service remain critical elements for all successful operations, the effective use of advanced technology plays an ever-increasing role in the proper management of a foodservice operation.

Many guests now use their mobile devices to choose a restaurant, place their orders online ahead of their actual arrival, and then use a **mobile wallet** to pay for their meals. Today, costs related to technology play an ever-important role in total cost management.

**Key Term**

**Mobile wallet:** A virtual wallet that stores payment card information on a mobile device. Mobile wallets provide a convenient way for a user to make in-store payments at merchants listed with the mobile wallet service provider. Also known as a digital wallet.

While the role of technology is ever-increasing in many areas, most foodservice operators find they must carefully manage their technology costs in the key areas of:

✓ Marketing
✓ Guest Services
✓ Operations Management

**Key Term**

**Proprietary website:** A website in which the foodservice operator controls all the website's content and can readily make changes to it.

**Marketing**

Today, every open-to-the-public foodservice operation must include a professionally designed and maintained **proprietary website** as a key part of its marketing efforts.

An effective proprietary website provides driving directions, phone numbers, menus, nutritional information, customer reviews, and more. The cost to create and properly maintain their websites is a marketing expense of increasing importance to all foodservice operators.

Similarly, the cost of maintaining a significant presence on social media is of ever greater importance. Fewer and fewer foodservice operators now rely on print media advertising to attract guests while social media attracts increasing numbers of tech-savvy guests. While Facebook, YouTube, WhatsApp, Instagram, and Yelp are currently examples of popular social media sites, effective operators must monitor these and emerging social media sites that might directly impact their businesses. Regardless of their decisions regarding the best social media sites on which to maintain a significant presence, the cost of maintaining that presence represents another important marketing cost that is classified as an "other expense."

---

**Technology at Work**

Foodservice operators who create a proprietary website must choose a web host. Currently, GoDaddy, Amazon Web Services, and Google Cloud* are the three largest web hosts in the United States. These may fit the needs of a specific foodservice operator and should be considered as potential web hosts. However, there are likely other web hosts that might better address the operator's needs.

The best web hosting decisions are made when operators know what their website should do. For example, will guests visiting the website be able to do online ordering and payment, or will they simply be viewing a menu in a print format? Similarly, will guests be able to make a reservation on the website or join and review guest loyalty program points?

---

*(Continued)*

The features and functions to be offered on a foodservice operation's website most often help determine the best web host at the best cost. To see a list of available foodservice-oriented web hosts that could be considered before making a final selection, enter "best web hosts for restaurants" in your favorite search engine and view the results.

*Ranking as of March 3, 2023: https://www.hostingadvice.com/how-to/largest-web-hosting-companies/

### Guest Services

While advances in technology allow guests to easily use an operation's website and view information about it on social media, guest ordering and payment methods are also affected. Software programs and apps for guest-owned smart devices now allow guests to easily place their orders online or with their mobile devices. These tools save time for both guests and operators. Similarly, advances in technology now readily allow guests in some operations to make their menu selections using self-ordering kiosks or terminals located on-site or at their tables.

In addition to changes in how guests place their orders, guests increasingly seek convenience in how they pay for their purchases. Tablet-based pay-at-the-table programs are just one example of this. The advent of electronic wallets and apps such as Apple Pay and Android Pay does provide convenience for guests, but they also come with charges to the operator, and these other expense costs must be carefully managed.

For some operators, one of the most significant guest service-related other expenses incurred are those of third-party delivery fees (see Chapter 7). To illustrate, if a guest used a third-party delivery app to purchase $100 worth of food from a foodservice operation, the operation might receive, for example, $70 from the delivery service and the service would keep $30 as their delivery fee.

In this scenario, the operation would record revenue (gross sales) of $100, and then record the $30 delivery charge in an appropriate account. While there currently is not complete agreement within the foodservice industry about which account is best used for the recording delivery fees (or if they should be recorded in their own designated account), in all cases these fees can be significant and must be documented if accurate income statements are to be produced at the end of each operation's accounting periods.

### What Would You Do? 9.1

"Well, I don't think it's complicated at all," said Amir, the assistant manager of the Hardrock Deli. "The delivery fees have to be considered a cost of sales. It's no different than the ingredients we use to make our sandwiches. Delivery fees are a cost we incur when we make a sale. So, we should record them as a cost of sales."

Amir was talking to Peggy Richards, the owner of the Hardrock Deli, a popular New York-style delicatessen located near a major college campus, and with Anna Rodgers, the deli's manager.

Peggy's restaurant features traditional deli fare like corned beef and pastrami sandwiches, homestyle soups, and traditional cheesecakes. Peggy had just entered an agreement with DineInDash, a third-party delivery service. The meeting taking place now was to discuss how to best record and account for the delivery fees that would be charged.

"Well, that doesn't make any sense to me," said Anna, "if we started our own delivery service, then we would charge the labor account for the cost of the people making the deliveries. So even though they're not our direct employees, I think the delivery fees have to be a labor cost."

"I don't know about that Anna," said Peggy, "if we delivered ourselves, we certainly wouldn't charge the cost of the vehicles and fuel to our labor account. Those costs would be placed in a different expense category for sure. And really, the reason I made the decision to go with DineInDash was to make more sales. Seems to me their fees should be a marketing cost. Guests will go to their website, see our information, and place their order. How is that any different than seeing an ad in a magazine or the campus newspaper and then coming to visit us? If we were going to record the cost of an ad, we would record it as a marketing cost, wouldn't we?"

**Assume you were Peggy. Would you agree with Amir, or with Anna? Or is your original thought to record the fees as a marketing cost still your position? Regardless of your decision, why would it be important to record these costs in the same manner every time an income statement for your operation was produced?**

### Operations Management

Advancements in technology have impacted virtually all areas of foodservice operations management, but most are not available for free. The specific costs incurred will vary by operation; however, a few common examples include:

1) Bluetooth temperature monitoring systems to help operators maintain proper food temperatures on a 24/7 basis and reduce the risk of foodborne illnesses.
2) Optimized schedule makers and scheduling software to help operators generate and distribute weekly work schedules that consider forecasted sales levels to all employees.

3) **Onboarding**, the process of orienting newly hired employees to an operation, can now be greatly sped up through the development and use of virtual reality video tours that allow new workers to "see" key aspects of an operation before they even begin to work in it.

4) Digital inventory tracking software that analyzes standardized recipes and POS sales data to create purchase order lists for management. These lists compare amounts on hand with forecasted needs to optimize product inventory levels.

5) POS-compatible software designed to take and manage guests' dining reservations, monitor **table turnover rates** and table wait times, and even suggest optimal seating arrangements.

6) Mobile ordering for menu item pick-up programs to enhance guest convenience. An increasing number of guests opt for take-out, rather than dine-in, experiences. Orders placed online for eventual customer pickup have, in some cases, created the need for "park-thru" lanes that permit guests to park close to a restaurant's entrance. This, in turn, enables guests to ease the process of picking up pre-ordered menu selections.

**Key Term**

**Onboarding:** The ways in which new employees acquire the necessary knowledge, skills, and behaviors to become productive members of a foodservice operation's staff.

**Key Term**

**Table turnover rate:** A measure of the amount of time a party occupies a dining table over a specific period of time.

For example, if an operation served 15 guests who occupied 5 tables between 7:00 p.m. to 11:00 p.m., the turnover rate would be calculated as:

15 guests/5 tables = 3 turns over a period of 4 hours.

All these operations management advancements (and many more) may be of great value. However, the costs of implementing each of them will always be categorized as one or more of an operation's "other expenses."

---

**Find Out More**

One important challenge faced by many foodservice operators is that of recruiting and retaining qualified staff members.

The costs incurred in advertising vacant positions, interviewing potential staff members, and training those who have been selected can be significant. These costs can be minimized when an operation provides a high-quality work environment that includes fair treatment of workers and the demonstration of concern for their well-being.

There are numerous strategies and tactics that can be implemented to develop a productive workforce. One good source of information to optimize the quality of the workforce while minimizing development costs is a useful book developed specifically for the foodservice industry.

To review the contents of this valuable employee recruitment and retention-related resource, enter "Managing Employees in Foodservice Operations by Hayes and Ninemeier" in your favorite search engine and view the results.

## Occupancy Costs

Occupancy costs (see Chapter 1) refer to those expenses incurred by a foodservice unit that are related to occupying and paying for its physical facility (building). For the foodservice operator who is not the building's owner, most occupancy costs will be non-controllable. Rent, mortgages, taxes, and interest on occupancy-related debt are real costs, but they are most often beyond the immediate control of a unit operator. However, for owners of a foodservice operation, occupancy costs are a primary determinant of profit on sales percentages and return on dollars invested.

If an operation's occupancy costs are too high because of unfavorable rent or lease arrangements or due to excessive debts and required loan repayments, the operator may face extreme difficulty in generating cash from operating profits. Food, beverage, labor, and other controllable operating expenses can only be managed to a point. After that, excessive efforts to reduce costs can result in decreased product quality, reduced guest service levels, and\or lower guest satisfaction inputs. If occupancy costs are unrealistically high, no amount of effective cost control can help to "save" the operation's profitability.

Total controllable and non-controllable other expenses in an operation can range from 10% to 25% (or even more) of the unit's total sales. Experienced operators know that, although other expenses are sometimes considered to be "minor" expenses, they are extremely important to overall operational profitability. This is especially true in a situation where the number of guests to be served is fixed, or nearly so, and the prices an operation is allowed to charge for its products are fixed as well. This is the case in many non-commercial operations such as business dining, health care, and university foodservices. In these and similar cases, an operator's abilities to control other expenses will be vital to the success of their business.

## Calculating Other Expense Costs

When assessing and managing other expenses, two control and monitoring calculations are very helpful to foodservice operators. These are:

1) Other expense cost percentage
2) Other expense cost per-guest

Each of the above tools can be used effectively in specific management situations, so it is important to understand and know how to use both calculations.

## Other Expense Cost Percentage

Other expenses can be analyzed based on their proportion (percentage) of total sales. The formula used to calculate a foodservice operation's other expense cost percentage is:

$$\frac{\text{Other Expense}}{\text{Total Sales}} = \text{Other Expense Cost}\%$$

The other expense cost percentage is a ratio (see Chapter 8) that can be useful for budgeting and comparison purposes. For example, if an operation with an advertising expense of $5,000 in a month served 10,000 guests and achieved total sales of $78,000 for that same month, its advertising cost percentage would be calculated as:

$$\frac{\$5,000}{\$78,000} = 0.064 \text{ or } 6.4\%$$

The calculation required to calculate an other expense cost percentage requires that the other expense cost category be divided by total sales. In many cases, this approach yields useful information. In some cases, however, this computation alone may not provide adequate information. Then using the concept of other expense cost per-guest can be very helpful.

## Other Expense Cost Per-guest

Some foodservice operators find it helpful to calculate their other expenses on a cost per-guest basis. The formula used to calculate a foodservice operation's other expense cost per-guest is:

$$\frac{\text{Other Expense}}{\text{Number of Guests Served}} = \text{Other Expense Cost Per-guest}$$

In a variety of situations, an operator's other expense cost per-guest provides more useful information than an other expense cost percentage. To illustrate, consider Scott, the co-owner/operator of Chez Scot, a fine-dining establishment in a suburban area of a major city.

One of Scott's major other expenses is linen. He uses both linen tablecloths and napkins. Scott's partner, Joshua, believes that linen costs are a variable operating expense and should be monitored by using a linen cost percentage. In fact, says

| Month | Total Sales | Linen Cost | Cost % |
|-------|-------------|------------|--------|
| January | $ 68,000 | $ 2,720 | 4.00% |
| February | $ 70,000 | $ 2,758 | 3.94% |
| March | $ 72,000 | $ 2,772 | 3.85% |
| April | $ 71,500 | $ 2,753 | 3.85% |
| May | $ 74,000 | $ 2,812 | 3.80% |
| Total | $355,500 | $13,815 | 3.89% |

**Figure 9.5** Chez Scot Linen Cost Percentage

Joshua, records indicate that the operation's linen cost percentage has been declining over the past five months; therefore, current linen cost control systems must be working well.

As shown in Figure 9.5, the operation's linen cost percentage has been declining over the past five months. Scott, however, is convinced that there are linen control problems. He has monitored linen expense on a cost per-guest basis. His information is presented in Figure 9.6, and it validates Scott's concern because there has been a control problem in the linen area.

While Figure 9.5 does indicate a declining linen expense percentage, Figure 9.6 clearly shows that the linen cost per-guest served has increased from a $1.06 low in January to a May high of $1.22.

Chez Scot is enjoying increased sales ($68,000 in January vs. $74,000 in May as shown in Figure 9.5). However, its guest count is declining (as shown in Figure 9.6, there were 2,566 guests in January and 2,305 guests in May). Since the guest count per month is declining, but sales have increased, the average sale per-guest (check average) has obviously increased. This is a good sign because it indicates that each guest is spending more.

| Month | Linen Cost | Number of Guests Served | Cost Per-guest |
|-------|------------|-------------------------|----------------|
| January | $ 2,720 | 2,566 | $1.06 |
| February | $ 2,758 | 2,508 | $1.10 |
| March | $ 2,772 | 2,410 | $1.15 |
| April | $ 2,753 | 2,333 | $1.18 |
| May | $ 2,812 | 2,305 | $1.22 |
| Total | $13,815 | 12,122 | $1.14 |

**Figure 9.6** Chez Scot Linen Cost Per-guest

The fact that fewer guests are being served per month should, however, result in a *decrease* in demand for linen and, thus, a decline in total linen cost (because it is a variable cost). In fact, on a per-guest basis, linen costs are up. Scott is correct to be concerned about possible problems in the area of linen cost control.

The other expense cost per-guest formula can be used any time management believes it can be helpful or when lack of a sales figure makes the regular computation of other expense cost percentage inappropriate.

The effective control of production and service costs, labor costs, and the other expense costs described in this chapter is critical to the success of all foodservice operations. It is also essential that foodservice operators price the menu items they sell in a way that allows them to cover these operating costs and still provide a reasonable profit level. The ability to properly price menu items is important, and it will be the sole topic of the next chapter.

---

**What Would You Do? 9.2**

"Wow, this is weird. I know it was hot last month, but I never expected our electric bill to look like this!" said Felix Unger, the manager of the Italian Garden restaurant.

Felix was talking to Myron Bollinger, the restaurant's owner.

Each month Felix and Myron reviewed the operation's P&L. The previous month had been a busy one, but the weather was unseasonably hot. As a result, the operation's electric bill was nearly 40% higher than the prior year. It was also much higher than the amount that had been budgeted for the month's utility bill.

Since the utility bill was so high, the restaurant's other expense percentage for utilities had cut into the operation's profit in a significant way.

"Well, it certainly was hot last month," said Myron. "And our costs were way up, but what can we really do about it?"

**Assume you were Felix. Where would you go to find information about reducing HVAC costs in particularly high-cost months? What specific steps might you take in the future to help reduce energy costs in your operation during very warm periods, and how would you best communicate these steps to your staff?**

---

## Key Terms

| | | |
|---|---|---|
| Standard operating procedure (SOP) | Maintenance log | Onboarding |
| | Mobile wallet | Table turnover rate |
| Kilowatt hour (kwh) | Proprietary website | |

## Operator's 10-Point Tactics for Success Checklist

Evaluate your need for, and the current status of, each of the following operational tactics. For those tactics you think are important, but not yet in place, develop an action plan for its implementation including who will be responsible for the tactic's completion and the target date by which it should be completed.

| Tactic | Don't Agree (Not Done) | Agree (Done) | Agree (Not Done) | If Not Done | |
|---|---|---|---|---|---|
| | | | | Who Is Responsible? | Target Completion Date |
| 1) Operator understands profits can be increased by controlling other expenses. | ____ | ____ | ____ | | |
| 2) Operator recognizes those characteristics of other expenses that make them controllable. | ____ | ____ | ____ | | |
| 3) Operator recognizes those characteristics of other expenses that make them non-controllable. | ____ | ____ | ____ | | |
| 4) Operator can classify other expenses as being either fixed, variable, or mixed. | ____ | ____ | ____ | | |
| 5) Operator understands the importance of monitoring and controlling other expenses in the food and beverage operations. | ____ | ____ | ____ | | |
| 6) Operator understands the importance of monitoring and controlling other expenses in equipment maintenance and repair. | ____ | ____ | ____ | | |
| 7) Operator understands the importance of monitoring and controlling other expenses in technology-related equipment and services. | ____ | ____ | ____ | | |

*(Continued)*

| Tactic | Don't Agree (Not Done) | Agree (Done) | Agree (Not Done) | If Not Done | |
|---|---|---|---|---|---|
| | | | | Who Is Responsible? | Target Completion Date |
| 8) Operator understands the importance of monitoring and controlling other expenses in occupancy costs. | ——— | ——— | ——— | | |
| 9) Operator can calculate other expense cost as a percentage of total sales. | ——— | ——— | ——— | | |
| 10) Operator can calculate other expenses on a cost per-guest basis. | ——— | ——— | ——— | | |

# 10

# Profitable Pricing

**What You Will Learn**

1) The Relationship Between Foodservice Prices and Profits
2) The Factors Affecting Foodservice Prices
3) The Methods Used to Price Menu Items
4) How to Evaluate Pricing Efforts in Foodservice Operations

**Operator's Brief**

In this chapter, you will learn that properly pricing menu items is an important skill for all foodservice operators even though prices are viewed differently by sellers and buyers. For operators, costs and profits are critical when establishing menu prices, but guests are concerned about the value they receive. Therefore, in addition to pricing menu items to be profitable, you must also ensure menu prices communicate real value to guests.

Numerous factors influence the prices you will charge for your menu items. Among the most important are economic conditions, competition, level of service, type of guest, and product quality and costs. Additional pricing factors include portion size, delivery style, meal period, location, and bundling: a pricing strategy that combines multiple menu items that are then sold at a price lower than that of the bundled items when purchased separately. In this chapter, you will learn about all these important factors.

When establishing menu prices, some operators use a product cost-based approach. They believe a menu item's production cost relative to its price is of most concern. When using this pricing approach, menu items with lower product cost ratios are thought to be more desirable to sell than those with higher product cost ratios. Other operators use a more profit-oriented approach to

*(Continued)*

establish their menu prices. Menu items providing high profit-per-sale are considered more desirable to sell than those with lower profit levels.

Regardless of the pricing approach you elect, it is important to regularly evaluate menu items that are the most popular and profitable. Then, you can modify or even eliminate poor selling or unprofitable items and better promote those that are the most popular and most profitable. In this chapter, you will learn how to do these things.

---

**CHAPTER OUTLINE**

Pricing for Profits
    The Importance of Price
    The Operator's View of Price
    The Guest's View of Price
Factors Affecting Menu Pricing
    Economic Conditions
    Local Competition
    Level of Service
    Type of Guest
    Product Quality
    Unique Alternatives
    Portion Size
    Delivery Method
    Meal Period
    Location
    Bundling
Methods of Food and Beverage Pricing
    Cost-based Pricing
    Contribution Margin-based Pricing
Evaluation of Pricing Efforts
    Menu Engineering
    Menu Modifications

---

# Pricing for Profits

Foodservice operators must price their menu items to help ensure the long-term profitability of their businesses. Experienced foodservice operators know that, if menu prices are too low, an operation may be popular but not profitable. Alternatively, if menu prices are too high, the popularity of a foodservice operation will likely suffer because few guests will regularly visit the operation. As a result, operators who establish menu item prices must understand how to do so effectively.

## The Importance of Price

**Price** plays a large role in the profitability of a foodservice operation. To best understand the importance of price, foodservice operators must understand that the term as used in the foodservice industry has two separate definitions.

Note that, in both uses (noun or verb), the concept of an *exchange* between a buyer and a seller is important. The foodservice operator gives up (exchanges) a menu item as it is purchased, and the foodservice guest gives up (exchanges) the item's selling price for the menu item.

Foodservice operators can typically charge any prices they want to charge. However, potential guests can accept or reject the operator's opinion that prices charged are fair and will provide good value to them (the guests). In fact, a foodservice operation itself can be selected by a foodservice guest based primarily on the prices charged and the guest's perceptions of those prices.

**Key Term**

**Price (Noun):** A measure of the value given up (exchanged) by a buyer and a seller in a business transaction.

For example: "The price of the chicken sandwich combo meal is $9.95."

**Price (Verb):** To establish the value to be given up (exchanged) by a buyer and a seller in a business transaction.

For example: "We need to price the chicken sandwich combo meal."

To best understand pricing in the foodservice industry, it is important to first recognize that price is viewed very differently from the perspectives of an operator (the seller) and a guest (the buyer).

## The Operator's View of Price

In most cases, foodservice operators can decide what products to sell, where and when they will sell them, and how they communicate their selling prices to their potential guests. Foodservice operators can propose their prices, but they also face the possibility that potential guests may *not* support their **value proposition**.

In the foodservice industry, price and value are closely related concepts. It is often stated that the value of any item is equal to what a buyer

**Key Term**

**Value proposition:** A statement that clearly identifies the benefits an operation's products and services will deliver to its guests.

will pay for it. If this is true, when a sale is *not* made, the buyer either believed the item was not worth the asking price or a lower cost alternative was available that was also considered worth its asking price.

From the perspective of most operators, a fair selling price should be an amount equal to the operator's incurred costs plus a reasonable desired profit.

Stated mathematically, that concept is:

Item Cost + Desired Profit = Selling Price

When calculating menu item costs, operators must consider the estimated costs of food, beverage, labor, and all other costs required to operate their businesses. Note: Professional foodservice operators can generally calculate these costs quite accurately.

The question of "What is a reasonable desired profit?" is more subjective. In the foodservice industry, an operation's **profit margin** is the percentage of each buyer's revenue dollar the operation retains as profit. The higher the profit margin, the more profitable is the operation.

In most cases, a foodservice operator wants to generate a reasonable profit margin because doing so is critical to staying in business and receiving a fair return for the investment risks in the business. Therefore, consideration of operating costs incurred and the profit desired are the two most important concerns as operators establish their menu prices.

**Key Term**

**Profit margin:** The amount by which revenue in a foodservice operation exceeds its operating and other costs.

### The Guest's View of Price

All guests desire a good **value** for the products and services they purchase. If they do not feel they have received good value, the guests are unlikely to be satisfied and will be unlikely to return to the business. Foodservice operators must remember that, in any business transaction, value is determined by the buyer, not the seller.

For example, a foodservice operator may sincerely believe that an item's quality, quantity, and delivery method will provide good value to guests if it is sold for $19.95. If, however, too few customers share that view of value, the menu item will not be frequently sold. Stated mathematically, a buyer's perception of value is expressed as:

**Key Term**

**Value:** The amount paid for a product or service compared to the buyer's view of what they receive in return for the purchase price.

Buyer's Perceived Benefit(s) − Price = Value

When making a purchase, buyers want to receive *more* value than the value of what they are giving up. Stated another way, there are three possible buyer reactions to any seller's proposed selling prices as shown in Figure 10.1.

| Buyer's Assessment | Purchase Decision |
|---|---|
| 1) Perceived Benefit – Price = A value less than "0" | Do not buy |
| 2) Perceived Benefit – Price = A value equal to "0" | Do not buy in most cases |
| 3) Perceived Benefit – Price = A value greater than "0" | Buy |

**Figure 10.1** Buyer's Assessment of a Seller's Value Proposition

When buyers believe the benefit they will receive is *less than* zero (less than what they give up in exchange), they generally will not buy. When the benefit is *greater than* zero, guests are very likely to buy. When the perceived benefit is *equal to* zero, buyers are often indifferent to the purchase. In this scenario, if no other alternatives are available, they may make a purchase decision. If alternatives are available, buyers will likely consider these alternatives before making purchase decisions.

To best understand how buyers' perceived benefit assessments directly impact purchasing decisions, operators must first understand the concept of **consumer rationality**.

Consumer rationality assumes that buyers consistently exhibit reasonable and purposeful behavior. That is, buyers generally make purchase decisions based solely on their belief that it benefits them to do so.

**Key Term**

**Consumer rationality:** The tendency to make buying decisions based on the belief that the decisions are of personal benefit.

For foodservice operators, the acceptance of the concept of consumer rationality involves a willingness to look beyond the obvious and then attempt to understand exactly how buyers believe they will benefit from a business transaction.

In some cases, this is not easy. First, foodservice operators must resist the temptation to declare their guests are irrational (e.g., when operators criticize guests who state the operator's prices are "too high!").

All buyers like low prices, but what they seek most is value. Most foodservice guests are indifferent to the actual costs of operating a foodservice business and, therefore, they are indifferent to an operator's profit margin.

Also, rational buyers *do not* automatically equate a seller's price with the amount of value they (the buyers) will receive in an exchange. In fact, conventional wisdom advises them not to do so. From common sense and even from a legal perspective, buyers assessing a seller's value proposition are cautioned not to trust sellers. As a result, *caveat emptor,* the Latin phrase for "let the buyer beware," is known and well-understood by most consumers.

Since many buyers may be skeptical about a seller's initial value proposition, the foodservice operator's responsibility when pricing menu items is to ensure guests understand the answers to questions like "What do I get?" and "Why is it of value?" just as much as they understand the actual prices they will pay. Only then can buyers, who are increasingly sophisticated and web-savvy consumers, learn they will consistently receive *more* than the worth of the money they must pay when making their purchase decisions.

A foodservice operator's primary motivation to recover operating costs and generate a profit is very different from their guests' motivation to optimize the value received for the prices they pay. Therefore, while operators are wise to be concerned about their costs and profits, what they should be *more* concerned about is utilizing selling price as a means of communicating excellent value to their guests.

## Factors Affecting Menu Pricing

When foodservice operators find that profits are too low, they frequently question whether their prices (and therefore their revenues) are too low. Remember that the terms "revenue" and "price" are not the same thing. "Revenue" is the amount of money spent by all guests, and "price" refers to the amount charged for one menu item. Total revenue for that menu item is generated by the following formula:

Selling Price × Number Sold = Total Revenue

There are two components of total revenue. Price is one component, and the other is the number of items sold. Note: Generally (but not always!), the economic **law of demand** indicates that, as an item's selling price increases, the number of that item sold will decrease. Also, when an item's selling price decreases, the number of that item sold will most often increase.

Experienced foodservice managers know that increasing prices without giving added value to guests result in higher prices but (frequently) lower total revenue because there is a reduction in the number of guest purchases that are made. For this reason, menu prices must be evaluated because they can directly affect the prices

**Key Term**

**Law of demand:** The law of demand holds that the demand level (number of units sold) for a product or service declines as its price rises and increases as the price declines.

operators charge for the menu items they sell. Among the most important variables to analyze are:

✓ Economic conditions
✓ Local competition
✓ Level of service
✓ Type of guest
✓ Product quality
✓ Unique alternatives
✓ Portion size
✓ Delivery style
✓ Meal period
✓ Location
✓ Bundling

## Economic Conditions

The economic conditions existing in a local area or even the entire country can significantly impact prices operators charge for menu items. For example, a robust and growing local economy generally enables foodservice operators to charge higher prices for the items they sell. In contrast, a local economy in recession and/ or weakened by other events can limit an operator's ability to raise or even maintain current prices in response to rising product costs.

In most cases, foodservice operators cannot directly influence the strength of their local economies. They can and should, however, monitor local economic conditions and carefully consider the impact of these conditions when establishing their menu prices.

## Local Competition

The prices charged by an operation's competitors can be important, but this factor is sometimes too closely monitored by some foodservice operators. It may seem to some operators that their average guest is only concerned with low prices and nothing more. However, small price variations generally make little difference in the buying behavior of the average guest.

For example, if a group of young professionals goes out for pizza and beer after work, the major determinant is not likely whether the selling price for a small pizza is $17.95 in one operation or $19.95 in another. Other factors including quality, location, and parking availability may become important.

The selling prices of potential competitors are of concern when establishing a selling price, but experienced operators understand that a specific operation can always sell a product of lower quality for a lesser price. While competitors' prices can help an operator arrive at their property's own selling prices, it should not be the only determining factor.

The most successful foodservice operators focus on building guest value in their own operations and not on attempting to mimic their competitors' efforts. Even though operators may believe their guests only want low prices, it is important to remember that consumers often associate higher prices with higher quality products and, therefore, products that provide a better price/value relationship.

### Level of Service

The service levels an operation provides directly affects the prices the operation can charge. Many guests expect to pay more for the same product when service levels are higher. For example, a can of soda sold from a vending machine is generally less expensive than a similar sized soda served by a service staff member in a sit-down restaurant.

Service levels can impact pricing directly and, as the personal level of service increases, selling prices may also increase. Personal service ranges from the delivery of products to a guest's home to the decision to quicken service by increasing the number of servers in a busy dining room (which improves service quality by reducing the number of guests each server assists).

These examples should not imply that extra income from increased menu prices is necessary to pay for extra labor required to increase service levels. Guests are willing to pay more for increased service levels. However, higher prices must cover extra labor costs and provide extra profit as well. Many hospitality operators can survive and thrive over the years because of an uncompromising commitment to high levels of guest service, and they can charge menu prices reflecting their enhanced service levels.

### Type of Guest

All guests want good value for the money they spend. However, some guests are less price sensitive than others, and the definition of what represents good value can vary by the clientele served. Consider the pricing and purchasing decisions of convenience store customers across the United States. In these facilities, food products such as pre-made sandwiches, fruit, drinks, cookies, and other items are often sold at relatively high prices. The customers in these stores most desire speed and convenience, and they are willing to pay premium prices for their purchases.

Similarly, guests at a fine-dining steakhouse restaurant are less likely to respond negatively to small variations in drink prices than are guests at a neighborhood tavern. A thorough understanding of exactly who the potential guests are and what they value most is critical to the ongoing success of foodservice operators as menu prices are established.

## Product Quality

A guest's quality perception of a menu item can range from very low to very high, and perceptions are most often the direct result of how guests view an operation's menu offerings. These perceptions are directly affected by a menu item's quality, and they should never be shaped by the guest's view of an item's wholesomeness or safety. All foods offered by a foodservice operation must be wholesome and safe to eat. Guests' perceptions of quality are based on numerous factors of which only one is the quality of actual raw ingredients. Visual presentation, stated or implied ingredient quality, portion size, and service level are additional factors that impact a guest's view of overall product quality.

To illustrate, consider that, when most foodservice guests think of a "hamburger," they think of a range of products. A "hamburger" may include a rather small burger patty placed on a regular bun, wrapped in waxed paper, served in a sack, and delivered through a drive-through window. If so, guests' expectations of this hamburger's selling price will likely be low.

If, however, the guests think about an 8-ounce "**Wagyu beef** burger" with avocado slices and alfalfa sprouts on a fresh-baked, toasted, and whole-grain bun and served for lunch in a white-tablecloth restaurant, the purchase price expectations of the guests will be much higher.

A foodservice operator can select from a variety of quality levels and delivery methods when developing product specifications, and as they plan their menus and establish their prices. The decisions they make will have a direct impact on menu pricing.

> **Key Term**
>
> **Wagyu beef:** Beef from a Japanese breed of cattle that is highly prized for its marbling and flavor. In the Japanese language, "Wa" means Japanese, and "gyu" means cow.

For example, if a bar operator selects an inexpensive Bourbon to make whiskey drinks, he or she will likely charge less for whiskey drinks than another operator selecting a better (more expensive) brand. Guest perceptions of the value received from those buying the lower-cost whiskey drinks will likely be lesser than guests served a higher-quality product.

To be successful, foodservice operators should select the product quality levels that best represent the anticipated desires of their **target market** and their operations' own pricing and profit goals.

> **Key Term**
>
> **Target market:** The group of people with one or more shared characteristics that an operation has identified as most likely customers for its products and services.

## Unique Alternatives

Property operators should understand that the entire experience of the guests can be part of the value received as part of the menu pricing decision. Most guests agree that they would pay more for the views presented by the dining room. Think about an ocean view and even lights that reflect these scenes after dark, a local entertainer who performs regularly at the property, and/or numerous examples of property support amenities such as artwork and museum pieces that justify the history of the operation. The "value equation" really often considers more than just the food that is purchased!

## Portion Size

Portion size can often play a large role in determining a menu item's price. Great chefs are fond of saying that people "Eat with their eyes first!" This concept relates to presenting food that is visually appealing, and it also impacts portion size and pricing.

A pasta entrée filling an 8-inch plate may be lost on an 11-inch plate. However, guests receiving the pasta entrée on an 8-inch plate will likely perceive higher levels of value than those receiving the same entrée on an 11-inch plate, even though the portion size and cost to the operator in both cases may be almost identical.

Portion size is a function of both product quantity *and* presentation. Many successful cafeteria chains use smaller-than-average dishes to plate their food. For their guests, the image of price to value when dishes appear full comes across loud and clear.

In some foodservice operations, and particularly in those that are "all-you-care-to-eat" facilities, the previously mentioned principle again holds true. The proper dish size is just as critical as the proper sized scoop or ladle when serving the food. Of course, in a traditional table service operation, an operator must carefully control portion sizes because the larger the portion size, the higher the product costs.

Many of today's health-conscious consumers prefer lighter food with more fruit and vegetable choices. The portion sizes of these items can often be boosted at a low increase in cost. At the same time, average beverage sizes are increasing as are portion sizes of many side items such as French fries. If these items are lower-cost items, this can be good news for the operator. However, it is still important to consider the costs of larger portion sizes.

Every menu item to be priced should be analyzed to help determine if the quantity (portion size) being served is the "optimum" quantity. Operators would, of course, like to serve that amount, but no more. For operators managing their cost control efforts, the effect of portion size on menu prices is significant, and back-of-house staff should establish and maintain control over desired portion sizes.

**Find Out More**

Foodservice operators have a variety of choices when selecting software programs designed to help calculate the cost and pricing of their various menu items. These menu management programs allow operators to insert their standardized recipes, portion sizes, and the cost of the ingredients used to make the recipes, and then the items' portion costs are automatically calculated.

To review software systems and tools designed to help operators better monitor their portion costs in real time, enter "ingredient cost management systems for restaurants" in your favorite search engine and review the results.

## Delivery Method

Delivery methods have become an increasingly important factor in establishing menu prices. There are essentially four ways foodservice operators can deliver purchased menu items to their guests:

1) Dine-in service: Menu items are delivered to guests at their on-site tables or other seating areas.
2) Pick up/carryout: Guests receive their menu items from drive-through windows or pick up ordered items from their designated on-site carryout areas.
3) Operator direct delivery: Menu items are delivered to guests by an operation's own delivery employees.
4) Third-party delivery: Menu items are delivered to guests by third-party delivery (see Chapter 7) partners selected by the operation.

The delivery style with the greatest impact on menu prices is third-party delivery. From the perspective of guests, the decision to utilize a third-party delivery company such as Grubhub, DoorDash, or Uber Eats means the guest is placing important value on convenience as well as on the menu items they choose.

When guests order directly from a third-party delivery company, they pay for:

✓ The selling price of their selected menu items
✓ A service fee charged by the delivery company for providing the service
✓ A delivery fee for the food that is delivered
✓ A gratuity; an optional tip for the delivery driver

The COVID-19 pandemic era saw explosive growth in third-party delivery companies as many restaurants either closed indoor dining areas or severely restricted their inside seating capacities. Third-party delivery services are popular with many guests; however, the services of third-party delivery companies are not free to restaurants.

Depending upon the specific arrangement made between a foodservice operation and its third-party delivery partner(s), the operation will pay between 10% and 30% of a guest's total bill to the third-party delivery company. For example, if a foodservice operation has a customer who utilizes a third-party delivery app and purchases $100 worth of menu items from the operation, the operation will receive only $70.00 to $90.00 from the third-party delivery company.

Astute readers recognize that the foodservice operation pays a significant fee to satisfy their customers' desire for convenience. Many foodservice operators charge different (higher) menu prices when items are ordered from a third-party delivery app, or they avoid the use of third-party delivery services entirely.

## Meal Period

In some cases, guests willingly pay more for an item served in the evening than for that same item served during a lunch period. Sometimes, this is the result of a smaller "luncheon" portion size or some different menu items. However, in other cases, the portion size and service levels may be the same in the evening as earlier in the day. This is true, for example, in buffet restaurants that charge a different price for lunch than for dinner. Perhaps operators expect those on lunch break to spend less time in the operation, and they will then eat less. Alternatively, they may believe their guests simply seek to spend less for lunch than dinner.

Foodservice operators must exercise caution in this area. Guests should clearly understand why a menu item's price changes with the time of day. If this cannot be answered to the guest's satisfaction, it may not be wise to implement a time-sensitive pricing structure.

## Location

Some foodservice operators make their locations a major part of their key marketing messages, and location can be a major factor in price determination. This is illustrated, for example, by food facilities operated in themed amusement parks, movie theaters, and sports arenas.

Foodservice operators in these locations can charge premium prices because they have a monopoly on food sold to visitors. The only all-night restaurant on an interstate highway exit is in much the same situation. Contrast that with an operator who is just one of 10 similar seafood restaurants on a **restaurant row** in a seaside

**Key Term**

**Restaurant row:** A street or region well-known for having multiple foodservice operations in a close distance from each other.

resort town. In this case, it is unlikely that one of the 10 operations can charge prices significantly higher than its competitors based solely on location.

Operators should not discount the value of an excellent restaurant location, and location alone can influence price in some cases. Location does not, however, guarantee long-term success. Location can be an asset or a liability. If it is an asset, menu prices may reflect that fact. If location is a liability, menu prices may need to be lower to attract a sufficient clientele to ensure the operation achieves its total revenue and profit goals.

## Bundling

**Bundling** refers to the practice of selecting specific menu items and pricing them as a group (bundle) so the single menu price of all the items purchased together is lower than if the items in the group were purchased individually.

A common example of bundling is the combination meals (combo-meals) offered by quick-service restaurants. In many cases, these bundled meals consist of a sandwich, French fries, and a drink. Bundled meals, often promoted as "combo meals" or "value meals," are typically identified by a number (e.g., Number 6 or Number 7) or a single name, for example, Everyday Value Meal (Arby's), Mix and Match (Burger King), and Cheeseburger Combo (McDonald's), for ease of ordering.

Bundled menu offerings are carefully designed to encourage guests to buy all menu items included in the bundle, rather than to separately purchase only one or two of the items. Bundled meals are typically priced very competitively, and a strong value perception can be established in the guest's mind.

### Key Term

**Bundling:** A pricing strategy that combines multiple menu items into a grouping that is then sold at a price lower than that of the bundled items purchased separately.

---

**Find Out More**

You have now been introduced to several factors that can influence the prices foodservice operators charge for the items they sell. In the future, Leadership in Energy and Environmental Design (LEED) certification achieved by an operation may well constitute another such factor.

The LEED rating system developed by the U.S. Green Building Council (USGBC) evaluates facilities on several standards. The rating system considers sustainability, water use efficiency, energy usage, air quality, construction and materials, and innovation.

*(Continued)*

Increasingly, many consumers are willing to pay *more* to dine in LEED-certified operations. In addition, LEED-certified buildings are healthier for workers and for diners. The LEED certification creates benefits for foodservice operators, employees, and guests. It will likely continue to be of increasing importance to guests.

To learn more about LEED certification in the foodservice industry, enter "LEED-certified restaurant standards" in your favorite search engine and view the results.

**What Would You Do? 10.1**

"We have to lower our prices because there is nothing else we can do!" said Ralph, Director of Operations for the seven-unit Boston's Sub Shops. Boston's was known for its modestly priced, but very high-quality, sandwiches and soups.

Business and profits were good, but now Ralph and Rachel, who was one of the company's store managers, were discussing the new $8.99 "Foot Long Deal" sandwich promotion just announced by their major competitor: an extremely large chain of sub shops that operates thousands of units nationally and internationally.

"They just decided to lower their prices to appeal to value-conscious customers," said Ralph.

"But how can they do that and still make money?" asked Rachel.

"There's always a less expensive variety of ham and cheese on the market," replied Ralph. "They use lower-quality ingredients than we do, and we charge $10.99 for our foot-long sub. That wasn't bad when they sold theirs at $9.99. Our customers know we are worth the extra dollar. Now they are running their special at $8.99 . . . I don't know, but this might really hurt us," said Ralph, "What should we do?"

**Assume you were Rachel. How do you think your guests will respond to this competitor's new pricing strategy? What specific steps would you recommend to Ralph and Rachel that can help Boston's Subs address this new pricing/cost challenge?**

## Methods of Food and Beverage Pricing

Many factors impact how a commercial foodservice operation establishes its prices, and the methods used are often as varied as the operators who use the methods.

Menu item prices are directly affected by one or more of the factors previously described. However, in most cases, menu prices have historically been determined based on either an operation's costs or desired profit level per menu item. This makes sense when the operator's perspectives about the price formula introduced earlier in this chapter are re-examined closely:

Item Cost + Desired Profit = Selling Price

When foodservice operators focus on item costs to establish prices, they recognize that, in nearly all cases, menu items that cost more to produce must be sold at prices higher than lower-cost items. The actual prices charged in a foodservice operation are primarily determined by its owner, perhaps with assistance from the operation's managers and production staff. However, those primarily responsible for controlling costs in an operation must have a good understanding of both the cost-based and profit-based approaches to pricing menu items.

## Cost-based Pricing

When using cost-based pricing, a foodservice operator calculates the cost of the ingredients required to produce the menu item being sold. The cost of the menu item includes any menu item accompaniments. For example, when a foodservice operation sells a dinner entree including a salad and dinner rolls at one cost, the entree cost must also include that of the salad and dressing, and dinner rolls and butter included with the entree.

In most cases, cost-based pricing is based on the idea that the cost of producing an item should be a predetermined percentage of the item's selling price. With this pricing system menu items with lower percentages of cost when compared to their selling prices are typically considered more desirable than menu items with higher percentage costs.

Recall from Chapter 6 that the formula used to compute a menu item's actual **food cost percentage** is:

$$\frac{\text{Item Food Cost}}{\text{Selling Price}} = \text{Food Cost Percentage}$$

To illustrate, if it costs an operator \$4.00 to purchase the ingredients needed for a menu item, and the item is sold for \$20.00, the item's food cost percentage is calculated as:

$$\frac{\$4.00 \text{ Item Food Cost}}{\$20.00 \text{ Selling Price}} = 0.20 \text{ or } 20\%$$

When operators utilize food cost percentage as a major factor in pricing their menu items, the cost of producing one portion of the menu item must be accurately determined. When the item sold is a single portion item (e.g., one New York strip steak), the cost of producing one portion is relatively straightforward.

However, when menu items are produced in multiple portions (e.g., a pan of lasagna containing 12 portions is prepared), operators must calculate their portion cost (see Chapter 6) based on the standardized recipe (see Chapter 1) used to produce the menu item.

The use of standardized recipes is critical for operators utilizing product cost as a primary determinant of menu prices. The cost of producing a menu item must be known, and it must be consistently the same if selling prices are based on product costs. When ingredient costs change the increased cost must also be used to calculate the new portion costs resulting from standardized recipes now containing higher cost ingredients. These new portion costs may (or may not) cause an operator to change menu prices, but the actual costs should be known.

While for-profit foodservice operations are concerned with food cost percentages, non-profit operations are typically more interested in the cost to serve each guest.

Examples of non-profit operations in which cost per meal is important to operators include military bases, prisons, hospitals, senior living facilities, schools and colleges, and large business organizations that offer foodservices to their employees.

Whether the individuals served are soldiers, inmates, patients, residents, students, or employees, calculating an operation's cost per meal is easy because it uses a variation of the basic food cost percentage formula:

$$\frac{\text{Cost of Food Sold}}{\text{Total Meals Served}} = \text{Cost Per Meal}$$

For example, assume a non-profit operation incurred $65,000 in cost of food sold during an accounting period. In the same accounting period, the operation served 10,000 meals. To calculate this operation's cost per meal, the cost per meal formula is applied.

In this example, it would be:

$$\frac{\$65,000 \text{ Cost of Food Sold}}{10,000 \text{ Meals Served}} = \$6.50 \text{ Cost Per Meal}$$

Whether managers are most interested in their cost of food percentage or in their cost per meal served, it is essential that they first accurately calculate the cost of food sold. When foodservice operators have established a target food (or beverage) cost percentage, they can use it to determine their menu item prices.

For example, if a menu item has a food cost of $8.00 and an operation's desired cost percentage is 40% (the food cost percentage in the approved operating budget for the time the menu is used), the following formula can determine the item's menu price:

$$\frac{\text{Food Cost of Menu Item}}{\text{Desired Food Cost \%}} = \text{Selling Price}$$

In this example:

$$\frac{\$8.00 \text{ Food Cost of Menu Item}}{40\% \text{ Desired Food Cost}} = \$20.00 \text{ Selling Price}$$

Another method of calculating selling prices based on predetermined product cost percentage goals uses a pricing factor (multiplier) assigned to each potentially desired food or beverage cost percentage. This factor, when multiplied by an item's portion cost indicates a selling price that yields the desired product cost percentage. Some of the commonly used pricing factors are presented in Figure 10.2.

The above pricing factor method of establishing menu prices is easy to use. For example, if an operator wants a 25% product cost, and a menu item has a food cost of $4.50, the following pricing formula would be used:

| Desired Product Cost % | Pricing Factor |
|---|---|
| 20 | 5.000 |
| 23 | 4.348 |
| 25 | 4.000 |
| 28 | 3.571 |
| 30 | 3.333 |
| 33⅓ | 3.000 |
| 35 | 2.857 |
| 38 | 2.632 |
| 40 | 2.500 |
| 43 | 2.326 |
| 45 | 2.222 |

**Figure 10.2**   Pricing Factor Table

$$\text{Item Food Cost} \times \text{Pricing Factor} = \text{Selling Price}$$

In this example, that would be:

$$\$4.50 \text{ Item Food Cost} \times 4.0 \text{ Pricing Factor} = \$18.00 \text{ Selling Price}$$

The two methods to determine a proposed selling price based on product cost percentage yield identical results. With either approach, an item's selling price is determined with the goal of achieving a specified food or beverage cost percentage for each item sold.

---

**Technology at Work**

Foodservice professionals will have no difficulty identifying numerous companies that offer menu pricing software. From single-unit diners to food trucks, wine bistros, fast food, fine dining and multi-unit concepts, every operator can benefit from menu pricing software that is both powerful and inexpensive.

To learn more about currently available menu pricing aids that are easy to use, enter "menu pricing software" in your favorite search engine and review the results.

## Contribution Margin-based Pricing

Some commercial foodservice operators use a profit-based, rather than cost-based, pricing method to establish menu item selling prices. These operators set their prices based on each of their menu item's **contribution margin (CM)**.

CM for a single menu item is defined as the amount of money that remains after the product cost of the menu item is subtracted from the

**Key Term**

**Contribution margin (CM):** The dollar amount remaining after subtracting a menu item's product costs from its selling price.

item's selling price. It is, then, the dollar amount that a menu item "contributes" to pay for labor and all other expenses and to contribute to an operation's profits.

To illustrate CM pricing, assume a menu item sells for $18.75 and the cost of the food to produce the item is $7.00. In this example, the CM for the menu item is calculated as:

Selling Price – Item Food Cost = Contribution Margin (CM)

or

$18.75 Selling Price – $7.00 Item Food Cost = $11.75 CM

When this approach is used, the formula for determining a menu item's selling price is:

Item Food Cost + Contribution Margin (CM) Desired = Selling Price

When using the CM approach to establish selling prices, operators most often develop different CM targets for various menu items or groups of items. For example, in an operation where items are priced separately, entrées might be priced with a CM of $10.50 each, desserts with a CM of $5.25 each, and non-alcoholic drinks with a CM of $3.75.

To apply the CM method of pricing, foodservice operators use a two-step process.

Step 1: Determine the average contribution margin required for each item.
Step 2: Add the contribution margin required to the item's product cost.

Step 1: Operators determine the average CM they require based on the number of items to be sold or on the number of guests to be served. The process used for each approach is identical.

For example, to calculate CM based on the number of items to be sold, operators add their non-food operating costs to the amount of profit they desire, and then divide the result by the number of items expected to be sold:

$$\frac{\text{Non-food Costs} + \text{Profit Desired}}{\text{Number of Items to Be Sold}} = \text{CM Desired Per Item}$$

To calculate CM based on the number of guests to be served, operators divide all non-food operating costs, plus the amount of profit they desire, by the number of expected guests:

$$\frac{\text{Non-food Costs} + \text{Profit Desired}}{\text{Number of Guests to Be Served}} = \text{CM Desired Per Guest Served}$$

For example, if an operator's budgeted non-food operating costs for an accounting period are $125,000, desired profit is $15,000, and the number of items estimated to be sold is 25,000, the operator's desired average CM per item would be calculated as:

$$\frac{\$125,000\,(\text{Non-food Costs}) + \$15,000}{25,000\,(\text{Number of Items to Be Sold})} = \$5.60\,\text{CM Desired Per Item}$$

Step 2: Operators complete this step by adding their desired CM per item (or guest) to the cost of preparing a menu item. For example, if (as in the example above) an operator's desired average CM per item is $5.60 and a specific menu item's food cost is $3.40, the item's selling price would be calculated as:

$$\$3.40\,\text{Item Food Cost} + \$5.60\,\text{CM Desired} = \$9.00\,\text{Selling Price}$$

The CM method of pricing is popular because it is easy to use, and it helps ensure that each menu item sold contributes to an operation's profits. When using CM to set menu prices, the prices charged for menu items vary only due to variations in product cost. When managers have accurate budget information about their non-food costs (taken from previous income statements) and realistic profit expectations, the use of the CM method of pricing can be very effective.

Operators who utilize the CM approach to pricing do so believing the average CM per item sold is a more important consideration in pricing decisions than the product cost percentage. The debate over the "best" pricing method for food and beverage products is likely to continue. However, all operators must view pricing as an important process, and its goal is to consider a desirable price/value relationship for guests.

In the final analysis, the customer eventually determines what an operation's sales will be for each menu item. Experienced operators know that sensitivity to required profit and to guests' needs, wants, and desires are the most critical components of an effective pricing strategy.

## Evaluation of Pricing Efforts

Regardless of the method used to establish their selling prices, foodservice operators should regularly evaluate the results of their pricing efforts, and many operators evaluate menu items based on two key characteristics: popularity and profitability. **Menu engineering** is a term popularly used to describe one method that addresses and examines these two variables.

### Menu Engineering

Operators using the menu engineering process desire to produce a menu that maximizes the menu's overall CM (defined previously as the amount operators have available to pay for labor and all other expenses and to contribute to profit). As a result, operators should be keenly focused on their menu's overall CM.

To use menu engineering, operators must sort their menu items by two variables:

1) Popularity (number of each item sold)
2) Weighted contribution margin

#### Calculating Menu Item Popularity (Number Sold)

To calculate the average popularity (number sold) of a menu item, operators must first determine the average number of all items sold:

$$\frac{\text{Total Number of Menu Items Sold}}{\text{Number of Menu Items Available}} = \text{Average Number Sold}$$

For example, if an operator sold 5,000 entrees during a specific period, and the operator's menu lists 10 different entree choices, the average popularity of the entrees sold during that time period would be calculated as:

$$\frac{5,000 \text{ Menu Items Sold}}{10 \text{ Menu Items Available}} = 500 \text{ Menu Item Average Popularity (Number Sold)}$$

When using menu engineering and applying the 500 average popularity, any menu item that sold *more* than 500 times during the analysis period is classified as "High" in popularity, and menu items selling *less* than 500 times would be classified as "Low" in popularity.

### Calculating Weighted Contribution Margin

To continue the menu engineering process operators, must also define the weighted average CM of their menu items. Some operators confuse averages (means) with weighted averages. However, the distinction between the two is important.

To use a simple example, assume an operator collected the following data and wanted to calculate the average size of the sale made in their operation over a 3-day reporting period.

| Week Day | Guests Served | Total Sales | Average Sale |
|---|---|---|---|
| Monday | 50 | $500 | $10.00 |
| Tuesday | 150 | $1,650 | $11.00 |
| Wednesday | 250 | $3,000 | $12.00 |
| Total/Average | 450 | $5,150 | ? |

In this example, to calculate the size of the "average" sale on Monday through Wednesday, one should NOT use the unweighted formula typically used to calculate a mean (average). That *unweighted formula* is:

$$\frac{\$10.00 + \$11.00 + \$12.00}{3\,days}\; \$11.00 \text{ Per Day Average}$$

In fact, what the operator really wants to learn when calculating the average sales for the three days is "How much did the average guest spend in my operation from Monday through Wednesday?"

The number of guests served each day varied, so the operator must determine a *weighted* average sale formula as shown below:

$$\frac{\$5,150\,(\text{Total Sales During All 3 Days})}{450\,(\text{Total Guests Served in All 3 Days})} = \$11.44 \text{ Weighted Average Sale Per Guest}$$

Note that the operator's average sale size resulting from using unweighted and weighted average formulas in this example differ.

Returning to menu engineering, to calculate the average **weighted contribution margin** for their menu items, operators must first calculate the total contribution margin generated by all items sold, and then divide by the number of items sold as shown in Figure 10.3.

Column A in Figure 10.3 lists the name of individual menu items on the menu. Column B lists the total number of each menu item that was sold. Note that the sales (popularity) of the items vary from a low of 190 sold (Item 10) to a high of 1,050 sold (Item 4).

**Key Term**

**Weighted contribution margin:** The contribution margin provided by all menu items divided by the total number of items sold.

Weighted contribution margin is calculated as:

$$\frac{\text{Total Contribution Margin of All Items Sold}}{\text{Total Number of Items Sold}} = \text{Weighted Contribution Margin}$$

Column C lists the individual CM of each of the 10 items offered for sale, and column D lists the total item CM generated by each menu item.

The value in Column D is calculated by multiplying the value in Column B times the value in Column C. In this example, the average number sold is 500, and the average weighted CM is $12.43 ($62,174 total item CM/5,000 sold = $12.43).

| Column A | Column B | Column C | Column D |
|---|---|---|---|
| Menu Item | Total Number Sold | Single Item Contribution Margin | Total Item Contribution Margin |
| 1 | 250 | $14.50 | $3,625 |
| 2 | 250 | $7.50 | $1,875 |
| 3 | 525 | $12.50 | $6,563 |
| 4 | 1,050 | $17.25 | $18,113 |
| 5 | 510 | $7.00 | $3,570 |
| 6 | 625 | $13.50 | $8,438 |
| 7 | 400 | $12.75 | $5,100 |
| 8 | 825 | $10.50 | $8,663 |
| 9 | 375 | $8.25 | $3,094 |
| 10 | 190 | $16.50 | $3,135 |
| Total | 5,000 | $120.25 | $62,174 |
| Average (Mean) | 500 | $12.03 | |
| Weighted contribution margin | | | $12.43 |

**Figure 10.3** Total Item Contribution Margin Worksheet for a 10-Item Menu (rounded to the nearest dollar)

| | | Popularity | |
|---|---|---|---|
| | | Low | High |
| **Contribution Margin** | High | High contribution margin<br>Low popularity<br><br>PUZZLE | High contribution margin<br>High popularity<br><br>STAR |
| | Low | Low contribution margin<br>Low popularity<br><br>DOG | Low contribution margin<br>High popularity<br><br>PLOW HORSE |

**Figure 10.4** Menu Engineering Matrix

After an operator has calculated the popularity and weighted contribution margin of the items listed on the menu, the items are sorted into a 2 × 2 menu engineering matrix containing four squares as shown in Figure 10.4.

Figure 10.4 shows that menu items with sales above the average level of popularity (500 sold in this example) are "High" in popularity, and items that sold less than 500 times are "Low" in popularity. Similarly, those menu items whose contribution margins are above the weighted contribution margin average ($12.43 in this example) are "High" in contribution margin, and items with a lower contribution margin are "Low" in contribution margin.

Many users of menu engineering name the items contained in the four squares for ease of remembering the characteristics of each item. These commonly used names (Puzzles, Stars, Dogs, and Plow Horses) are also shown in Figure 10.4.

Figure 10.5 shows where each of the 10 example menu items listed in Figure 10.3 would be located.

| | | Popularity | |
|---|---|---|---|
| | | Low | High |
| **Contribution Margin** | High | PUZZLE<br>Menu items 1, 7, and 10 | STAR<br>Menu items 3, 4, and 6 |
| | Low | DOG<br>Menu items 2 and 9 | PLOW HORSE<br>Menu items 5 and 8 |

**Figure 10.5** Menu Engineering Results

| Item | Characteristics | Problem | Marketing Strategy |
|---|---|---|---|
| Puzzle | High contribution margin, Low popularity | Marginal due to lack of sales | a) Relocate on the menu for greater visibility.<br>b) Consider reducing the selling price. |
| Star | High contribution margin, High popularity | None | a) Promote well.<br>b) Increase prominence on the menu. |
| Dog | Low contribution margin, Low popularity | Marginal due to low contribution margin and lack of sales | a) Remove from the menu.<br>b) Consider offering as a special occasionally, but at a higher menu price. |
| Plow Horse | Low contribution margin, High popularity | Marginal due to low contribution margin | a) Increase the menu price.<br>b) Reduce prominence on the menu.<br>c) Consider reducing portion size. |

**Figure 10.6**   Potential Menu Modifications

## Menu Modifications

Operators should regularly analyze their menus to make needed modifications and improvements. When using menu engineering, each menu item that fell within the four squares requires a special marketing strategy. Examples of these suggested menu modification strategies resulting from menu engineering are summarized in Figure 10.6.

---

**Technology at Work**

A foodservice operation's menu is much more than a list of foods and beverages. It can be a powerful marketing tool to familiarize customers with an operation. It can also get them excited about the unique items an operation offers for sale.

Regularly evaluating individual menu items for popularity and profitability is an important task because it helps identify both profit-producing items and those that perform poorly. With this information known, menus can be modified to optimize sales and profits.

Fortunately, there are several useful software programs available that help operators perform menu engineering analysis. To review the features and costs of such programs, enter "menu engineering software" in your favorite search engine and view the results.

One reason to perform menu engineering analysis is to identify items whose prices must or can be increased to enhance an operation's profitability. Some operators are hesitant to raise prices fearing customers will react negatively. Experienced foodservice operators, however, know that price is not the only determining factor when guests decide where to spend their dining-out dollars.

Quality of customer service, cleanliness, staff friendliness, and uniqueness of menu items offered are often *more* important than price. The best operators ensure all these aspects of their operations meet or exceed their guests' expectations. Then price increases acceptable to guests that help ensure an operation's long-term profitability can be implemented.

This chapter has indicated that foodservice operators base their prices on production costs and desired profit. Proper pricing produces the funds foodservice operators must acquire to pay their operating expenses. However, if funds are to be available to pay legitimate expenses, it is also important to safeguard the revenue a foodservice operation generates. Effective revenue control procedures begin at the time of guest payment and continue when a deposit is made in the operation's bank account. The proper control of revenue is so important it is the sole topic of the next chapter.

---

**What Would You Do? 10.2**

"$39.95—that's over ten dollars more than we charged for it yesterday!" said Shawn, the Dining Room manager at Chez Franco's restaurant.

Shawn was discussing the day's dinner menu with Aimée, the restaurant's executive chef. Aimée had just shown Shawn the daily menu insert his service staff would use that night. On the night's new menu, he noticed that the price of Lemon Red Snapper with Herb Butter, one of the operation's most popular dishes, had increased overnight. Yesterday, it sold for $27.95. Today, Aimée had priced it at $39.95.

"Tell me about it," replied Aimée, "our seafood supplier really raised the price on our latest delivery. My snapper cost is now up by almost $11.00 per pound. That's over $4.00 a portion. The supplier said there was a snapper shortage, and he wasn't sure how long it would last. With the new cost of snapper, I needed this price increase to keep our food cost percentage in line."

Shawn wasn't sure his servers or the guests to be served that night would be very happy with Aimée's pricing decision. The snapper was a very popular item, and that meant tonight lots of customers would likely notice the price increase and make a comment about it!

**Assume you were Shawn on the night this new menu price was initiated. How would you likely respond to a returning guest who questioned the significant price increase on the red snapper item? What do you think you should tell your servers to say in response to guests' anticipated reactions to this menu item's price increase? What else might you do to address this issue?**

## Key Terms

| | | |
|---|---|---|
| Price (noun) | Law of demand | Contribution |
| Price (verb) | Wagyu beef | margin (CM) |
| Value proposition | Target market | Menu engineering |
| Profit margin | Restaurant row | Weighted contribution |
| Value | Bundling | margin |
| Consumer rationality | | |

### Operator's 10-Point Tactics for Success Checklist

Evaluate your need for, and the current status of, each of the following operational tactics. For those tactics you think are important, but not yet in place, develop an action plan for its implementation including who will be responsible for the tactic's completion and the target date by which it should be completed.

| | | | | If Not Done | |
|---|---|---|---|---|---|
| Tactic | Don't Agree (Not Done) | Agree (Done) | Agree (Not Done) | Who Is Responsible? | Target Completion Date |
| 1) Operator understands the importance of price in the profitable operation of a foodservice business. | ____ | ____ | ____ | | |
| 2) Operator has carefully considered the difference between an operator's view of price and their guests' view of price. | ____ | ____ | ____ | | |
| 3) Operator has considered the impact of economic conditions, local competition, level of service, and guest type when establishing menu prices. | ____ | ____ | ____ | | |
| 4) Operator has considered the impact of portion size, delivery style, meal period, and location when establishing menu prices. | ____ | ____ | ____ | | |

| Tactic | Don't Agree (Not Done) | Agree (Done) | Agree (Not Done) | If Not Done Who Is Responsible? | Target Completion Date |
|---|---|---|---|---|---|
| 5) Operator understands the concept of bundling when establishing menu prices. | ___ | ___ | ___ | | |
| 6) Operator has considered the value of utilizing a product cost-based approach when establishing menu prices. | ___ | ___ | ___ | | |
| 7) Operator has considered the value of utilizing a contribution margin-based approach when establishing menu prices. | ___ | ___ | ___ | | |
| 8) Operator understands the importance of analyzing the menu as a means of better understanding guests' purchasing preferences. | ___ | ___ | ___ | | |
| 9) Operator understands how to use menu engineering to analyze their menus. | ___ | ___ | ___ | | |
| 10) Operator recognizes the importance of carefully implementing any menu price adjustments to minimize guest dissatisfaction and to optimize sales. | ___ | ___ | ___ | | |

# 11

# Cash and Revenue Control

## Operator's Brief

One of a foodservice operator's most important responsibilities is to protect business assets from theft and fraud. Foodservice operations can be an easy victim of these crimes, and control systems maintaining cash and other asset securities are essential.

In this chapter, you will learn about effective revenue control systems and how their components are developed, implemented, and monitored. A foodservice operation's cash assets are generated by guests' sales. Unfortunately, cash can be tempting to dishonest individuals, and its security is subject to external and internal threats that must be controlled.

In this chapter, you will learn important procedures for safeguarding cash from when it is received from guests until it is deposited in an operation's bank account. There are several objectives and useful strategies important as revenue security programs are developed, and this chapter explains them.

As revenue security is implemented and monitored, operators must pay special attention to five key activities. Each is detailed in this chapter, and you will learn why it is necessary to properly verify your operation's:

Receivable Revenue
Total Guest Charges
Sales Receipts
Sales (Bank) Deposits
Accounts Payable (AP) Accounts

---

**CHAPTER OUTLINE**

---

## The Importance of Revenue Control

Foodservice operators need an effective system to control cash revenues. These controls are typically important because many foodservice operations have hundreds (or more!) of daily cash sales transactions, and more than one employee typically handles the cash.

**Fraud** and **embezzlement** can occur all too frequently in some foodservice operations. Fortunately, there are several ways operators can reduce opportunities for these thefts to occur.

In most cases, there are three primary factors necessary for employee fraud and/or embezzlement to occur:

1) Need: Economic or psychological motives can encourage employees to steal. Some operators attempt to study their employees' lifestyles to determine whether selected staff members may steal. However, a better approach is to develop systems that reduce theft opportunities and that make it difficult for employees to "try and beat the system."

2) Failure of conscience: Sometimes thieves rationalize stealing to justify taking someone else's property. For example, some foodservice staff might believe they deserve more money because they are paid too little, and "stealing a little is okay because the business can afford it."

**Key Term**

**Fraud:** Deceitful conduct to manipulate someone into giving up something of value by (a) presenting as true something known by the fraudulent party to be false or (b) by concealing a fact from someone that may have saved him or her from being cheated.

**Key Term**

**Embezzlement:** A crime in which a person or entity intentionally misappropriates assets that have been entrusted to them.

3) Opportunity: Some otherwise honest employees may be tempted to commit theft when they are given the opportunity to do so. Foodservice operators can discourage theft by implementing systems and procedures that minimize opportunities for employees to steal.

Of the three factors noted above, foodservice operators have the most direct influence over theft opportunities. Therefore, cash and revenue control procedures should be designed to minimize these possibilities.

Most foodservice operations have some general operating characteristics that make them more vulnerable to theft than many other businesses. These include:

✓ Large numbers of individual cash transactions
✓ Items of relatively high value that are commonly used or available
✓ Workplace availability of products employees must otherwise purchase

Increasingly, foodservice guests use electronic payment forms rather than cash when paying their bills. However, cash is often exchanged between guests and employees. Cash banks made available to cashiers and servers can create potential theft risks as well.

In some operations, the use of relatively unskilled employees to work in low-paying positions can result in high employee turnover rates. These, in turn, can influence the internal control environment. Many foodservice operations are small, and even large operations may be organized into several small revenue outlets such as individual dining rooms, multiple bars, and, perhaps, take-out stations. These multiple sales locations can create strong needs for effective internal control systems that address both external and internal threats to revenue security.

### External Threats to Revenue Security

Foodservice guests can be threats to an operation's revenue, and an operation can lose revenue because some guests want to defraud it. This activity can take a variety of forms including that guests **walk** (skip) without paying their bill. This type of theft is less likely in a quick-service restaurant (QSR) or fast casual operation because payment is typically collected before, or at the same time, guests receive their menu items.

**Key Term**

**Walk (bill):** Guest theft that occurs when a guest consumes products that were ordered and leaves without paying the bill. Also referred to as a "skip."

In cases where a guest is in a busy table-service restaurant's dining room, it is easier for one or more persons in a dining party to leave without paying while their server is busy with other guests. In fact, it is sometimes easy for a guest or entire party to leave without settling their bill unless all staff members are

1) If guests order and consume their food before payment, servers can present the bill promptly when guests finish eating.
2) If the operation has a cashier in a central location in the dining area, he or she should be available and visible at all times.
3) If each server collects his or her own guest charges, he or she should return to the table promptly after presenting the guest's bill for payment.
4) Train employees to observe all exit doors near restrooms or other areas of the facility that may provide opportunities to exit dining areas without being easily seen.
5) If an employee sees a guest attempting to leave without paying the bill, he or she should notify the manager or person in charge immediately.
6) When approaching someone who has left without paying the bill, the manager should ask if the guest has inadvertently "forgotten" to pay. (In most cases, the guest will then pay the bill.)
7) If a guest still refuses to pay or flees the scene, the operator should make a written note of the following:
   a) Number of guests involved
   b) Amount of bill
   c) Physical description of guest(s)
   d) Vehicle description, if applicable, with license plate number when possible
   e) Time and date of the incident
   f) Name of the server(s) who served the guest(s)
8) If the guest is successful in fleeing the scene, police should be notified. Note: In no case should staff members, supervisors, or operators attempt to physically detain the fleeing guest(s). The reason: Liability is involved if an employee or guest is hurt, and this may result in a far greater loss than the value of a skipped food and beverage bill.

**Figure 11.1**   Steps to Reduce Guest Walks or Skips

vigilant. To help reduce this type of guest theft, implementation of the steps in Figure 11.1 can be helpful.

Another form of external theft can be used by the quick-change artist: a guest who attempts to confuse a cashier to give the guest excessive change. For example, a guest who should have received $5 in change may use a confusing routine to secure $15. To prevent this from happening, operators must train cashiers and instruct them to notify management immediately if there is suspicion of attempted fraud from quick-change routines.

## Internal Threats to Revenue Security

Most foodservice employees are honest, but some are not. In addition to protecting revenue from unscrupulous guests, operators must also be aware of employees who attempt to steal revenue from their properties.

Cash is the most readily usable asset in a foodservice operation, and it is often a major target for dishonest employees. In general, theft from service personnel does not occur by removing large sums of cash at one time because this method would normally be easy for operators to detect. Rather, dishonest service personnel may use numerous methods of removing a small amount of money at different times.

One common theft technique used by servers involves failure to record a guest's order in the operation's point-of-sale (POS) system. Instead, the server charges the guest and keeps the revenue from the sale. To reduce this possibility, operators must insist that all sales be recorded in the POS system and to match products sold to revenue received. (Note: This type of checking is often done as part of the cash-handling employee's check-out procedures.)

The POS system assigns a unique transaction number to the sale just as numbered guest checks did before the introduction of POS systems. A sale's transaction number (or a numbered guest check) is an electronic or handwritten record of what the guest purchased and how much the guest was charged for the item(s).

The use of electronic guest checks produced by the POS system is standard in the foodservice industry. A rule for all food and beverage production personnel is that no food or beverage item should be issued to a server unless the server first records the sale in the POS system. The items ordered are then displayed in the operation's production area. In some systems, the order may even be printed in the production area.

Increasingly, this method of entering guest orders can be accomplished in one step instead of two. Handheld and wireless at-the-table order-entry devices now allow servers (and guests in some cases!) to enter orders directly into an operation's POS system. This direct data entry system is fast, and it eliminates mistakes made when transferring handwritten guest orders to the POS system.

Regardless of how the order is created, kitchen and bar personnel should not issue any products to the server without this uniquely numbered sales transaction. When the guest wants to leave, the cashier (or server) retrieves the transaction and prepares a bill for payment by the guest. The bill includes the charges for all items ordered by the guest plus any service charges and taxes due, and the guest then pays the bill.

Another method of employee theft involves entering sales but failing to collect payment. To prevent this theft, management must have systems in place to identify **open checks** during and after each server's work shift.

Unless all open checks are ultimately presented to guests for payment and closed out, the value of menu items issued will not equal the money collected for those items' sale.

### Key Term

**Open check:** A guest check that initially authorizes product issues from the kitchen or bar but for which collection of the amount due has not been made (the account is not closed). Therefore, the amount due has not been added to the operation's sales total.

The totals of all transactions entered into the POS system during a predetermined time are electronically tallied, and operators can compare sales recorded by the POS system with the money in the cash drawer.

For example, a cashier working a shift from 7:00 a.m. to 3:00 p.m. might have recorded $1,000 in sales (including taxes) during that time. If that were the case and, if no errors in handling change occurred, the cash drawer should contain the $1,000 in sales revenue (in addition to the amount in the drawer at the beginning of the shift). If the drawer contains less than $1,000, it is said to be **short**; if it contains more than $1,000, it is said to be **over**.

Cashiers rarely steal large sums of cash from the cash drawer because this type of theft is easily detected. Operators should implement a policy to help ensure any significant cash shortages or overages will be investigated. Some operators believe that only cash shortages, not overages, need to be monitored, but this is not true.

Consistent cash shortages may be an indication of employee theft or carelessness and must be investigated. Cash overages, too, may result from sophisticated theft by the cashier. For example, assume a cashier defrauds an operation by removing and keeping $18.00 from a cash drawer but falsely reduces actual sales records by $20.00. The result is a $2.00 cash "overage!"

Even if operators implement controls to make internal theft difficult, the possibility of significant fraud still exists. Consequently, some food-service operators protect themselves from employee dishonesty by **bonding** their employees. Bonding involves purchasing an insurance policy against the possibility that employee(s) will steal.

When bonded, an employer can be covered for the loss of money or other property sustained through dishonest acts. Bonding can cover many acts including larceny, theft, embezzlement, forgery, misappropriation, or other fraudulent or dishonest acts committed by an employee alone or in **collusion** with others. Essentially, a business can select from several bonding options:

**Key Term**

**Short (cash bank):** A cashier's bank that contains less than the amount of cash, credit or debit card, or other authorized payment amount based on the number of menu items sold.

**Key Term**

**Over (cash bank):** A cashier's bank that contains more than the amount of cash, credit or debit card, or other authorized payment amount based on the number of menu items sold.

**Key Term**

**Bonding (employee):** Protection from loss resulting from a criminal event involving employee dishonesty and theft such as loss of money, securities, and other property.

**Key Term**

**Collusion:** Secret cooperation between two or more individuals that is intended to defraud an operation.

1) Individual—covers one employee (e.g., a foodservice operation's bookkeeper)
2) Position—covers all employees in a specific position (e.g., all bartenders and/ or all cashiers)
3) Blanket—covers all employees

If an employee has been bonded and an operator can determine that the employee was involved in the theft of a specific amount of money, the business will be reimbursed for all or part of the loss by the bonding company.

Although bonding will not eliminate all theft, it is a relatively inexpensive way to help ensure that an operation is protected from theft by employees who handle cash or other forms of operating revenue. Note: The bonding company will likely require detailed background information on employees before bonding them, and this is also an excellent pre-employment check to verify an employee's track record in prior jobs.

Operators must recognize that even good revenue control systems present the opportunity for theft if management is not vigilant, and this is especially so if two or more service employees work together to defraud the operation. Figure 11.2 identifies some common methods of internal theft involving service employees in a foodservice operation.

The scenarios addressed in Figure 11.2 do not include all possible methods of revenue loss. However, all operators must use an effective revenue security system to ensure all menu items sold generate sales revenue for the property.

---

1) Omits recording the guest's order and keeps the money the guest pays
2) Voids a sale in the POS system but keeps the money the guest paid
3) Enters another server's password in the POS system and keeps the money
4) Fails to finalize a sale (keeps a check open) and keeps the money
5) Charges guests for items not purchased and then keeps the overcharge
6) Changes the totals on payment card charges after the guest has left
7) Enters additional payment card charges and pockets the cash difference
8) Incorrectly adds legitimate charges to create a higher-than-appropriate total with the intent of keeping the overcharge
9) Purposely shortchanges guests when giving back change with the intent of keeping the extra change
10) Charges higher-than-authorized prices for products or services, records the proper price, and keeps the overcharge
11) Adds a coupon to the cash drawer and simultaneously removes sales revenue equal to the coupon's value
12) Declares a transaction to be complimentary (comped) after the guest has paid the bill
13) Engages in collusion to defraud the operation

---

**Figure 11.2** Common Methods of Theft by Service Employees

---

**Find Out More**

Unscrupulous foodservice employees can steal from their employers, but they also can also steal from an operation's guests.

In many cases, foodservice customers paying their bills hand over their credit cards to their servers, but they may not see where their card goes. This leaves opportunities for a guest's credit card information to be stolen on a credit-card skimming device that reads the magnetic stripe on a credit or debit card. When inserted into a card reader by a dishonest employee, the employee stores the card number, expiration date, and cardholder's name for later illegal use by the dishonest employee. Alternatively, an employee may also use a cell phone to photograph the front and back of a guest's credit card to obtain the name, credit card number, and the 3- or 4-digit security code number found on the back of the card.

Foodservice operators have a responsibility to help ensure the security of their guest's payments. Operators should remain updated about methods that could defraud their guests. To learn more about what can be done to help ensure the security of each guest's payment information, enter "preventing credit card theft by restaurant employees" in your favorite search engine and view the results.

---

**What Would You Do? 11.1**

"The beauty of our system," said Niles Carson, "is that you can monitor the actions of all your employees and supervisors."

Niles was talking to Gail Tuller, the owner of Fazziano's Italian Kitchen. Gail had called POS-Video Security, the company Phil represented because, for the second time this year, Gail had discovered a case of employee/supervisor collusion. Working together, the employee and supervisor stole revenue from their restaurant by manipulating their unit's POS system.

"I'm pretty sure I understand your system, but please go over it one more time," said Gail.

"Okay, Gail," said Niles. "Essentially, our new system goes beyond traditional surveillance methods by synchronizing the video being recorded with the data mined from your POS system to create detailed, customized video reports. Potentially fraudulent activity, such as manager (operator) overrides, coupons or comps, or even a cash drawer being open for too long, is tracked, and the corresponding video surveillance can be searched by transaction number. Data reports and streaming video, both real time and stored, can be accessed securely and remotely on a PC, smart device, or in the cloud."

*(Continued)*

"So, for example," replied Gail, "when a sales void occurs, your system identifies the portion of video that was recording at the time of the void and then allows me to view just that portion of the video so I can see what was happening in the restaurant during the transaction."

"Exactly," said Niles.

**Assume you were Gail. What types of employee fraud do you think could be uncovered using the technology offered by POS-Video Security's new product? Do you think the behavior of most dishonest cashiers and supervisors would change if they knew their actions were being video recorded? Explain your answer.**

## Developing a Revenue Security Program

Chapter 1 of this book introduced a foodservice operator's basic profit formula:

Revenue − Expenses = Profit

A close examination of the formula might lead some operators to think that 50% of their time should address managing and protecting revenue, and 50% of their time should consider managing expenses. This may be reasonable because all cost control systems will be of little use if they cannot initially collect the revenue their businesses generate, deposit that revenue into their bank accounts, and spend it only for legitimate expenses.

Errors in revenue collection or other asset security issues can result from simple employee mistakes or, sometimes, outright theft by guests or employees. An important part of every operator's cost control-related job is to devise revenue control systems to protect income. This is true regardless of whether the revenue is cash, checks, credit or debit card receipts, coupons, meal cards, or another guest payment method.

### Objectives of Internal Revenue Control Systems

The American Institute of Certified Public Accountants (AICPA) has a longstanding definition of internal control:

> *Internal control comprises the plan of organization and all of the coordinate methods and measures adopted within a business to safeguard its assets, check the accuracy and reliability of its accounting data, promote operational efficiency, and encourage adherence to prescribed managerial policies.*[1]

---

1 American Institute of Certified Public Accountants, Committee on Auditing Procedures, Internal Control-Elements of a Coordinated System and Its Importance to Management and the Independent Public Accountant (New York, 1949).

This definition indicates that internal control relies on an organization plan and use of methods and measures to attain four key objectives:

1) Safeguard assets
2) Check accuracy/reliability of accounting data
3) Promote operational efficiency
4) Encourage adherence to prescribed managerial policies

### Safeguard Assets

This objective addresses the protection of assets from losses including theft, resource maintenance (especially equipment) to ensure efficient utilization, and the safeguarding of resources (especially inventories) to prevent pilfering, waste, and spoilage. It also includes the safeguarding of cash receipts.

### Check Accuracy/Reliability of Accounting Data

This objective addresses the checks and balances within the accounting system designed to ensure the accuracy and reliability of financial information. Dependable data is needed for reports to owners, government agencies, and other outsiders, and it is also needed for an operator's internal use.

Foodservice operations of all sizes benefit from the use of the Uniform System of Accounts for Restaurants (USAR; see Chapter 1). Accounting information is most useful when it is timely, so internal control reports for operators must be prepared regularly and promptly.

### Promote Operational Efficiency

A foodservice operation's training program and procedures with effective supervision promote operational efficiency. For example, when cashiers are properly trained and supervised, resource control is optimized. Similarly, the use of mechanical and electronic equipment improves efficiency. For example, POS devices can relay orders from server stations to preparation areas and increase staff efficiency while recording all guest purchases.

### Encourage Adherence to Prescribed Managerial Policies

Asset security procedures should be designed to encourage compliance. For example, most foodservice operations have a policy that requires hourly paid employees to clock in and out personally; an employee cannot do this for another employee. Locating the time clock for easy managerial observation may encourage workers to follow the policy. In a similar manner, effective revenue control generally calls for assigned verification of cashier bank contents by at least two different individuals.

These four objectives may seem to conflict at times. For example, procedures for safeguarding an operation's assets may be so detailed that efficiency is reduced. Requiring multiple signatures to withdraw products from inventory may reduce theft, but the policy may require additional time. This, in turn, can increase labor costs that then could exceed potential inventory losses.

Perfect controls, even if possible, may not be cost-justified. Foodservice operators must evaluate the cost of implementing procedures against their benefits. The reason: An operation's internal control procedures are most efficient when costs and benefits are balanced.

## Elements of Internal Revenue Control Systems

Regardless of the specific foodservice operation, all effective revenue control systems have similar elements.

### Proper Leadership

Leadership is critical to the success of a foodservice operation's internal control system. Effective policies must be developed, clearly communicated, and consistently enforced. Each level of management must be responsible for ensuring that applicable control procedures are adequate, and that any exceptions to policies should be minimized and justified.

### Assigned Responsibility

When possible, responsibilities for a specific control activity should be assigned to one individual. Then that staff member can be given a set of standard operating procedures with the expectation they will be followed. When responsibility is given to one person, an operator knows where to start looking if a problem is identified.

For example, a cashier should be solely and fully responsible for a specific cash bank. To ensure this, no one except that employee should have access to the bank during the cashier's shift. In addition, there should be no sharing of the cash bank and/or responsibility for it.

### Separation of Duties

Separation of duties occurs when different personnel are assigned to accounting, asset responsibility, and production activities. Duties within the accounting function should also be separated. For example, different personnel should maintain an operation's **general ledger** and the operation's **cash receipts journal**.

The major objective of separating duties is to prevent and detect errors and theft. Unfortunately, it is not always possible to separate duties in very small foodservice operations. In these situations, operators must often assume multiple duties.

### Approval Procedures

Foodservice operators must authorize every business transaction. Authorization may be either general or specific. General authorization occurs when all employees must comply with selected procedures and policies when performing their jobs. For example, in all operations, servers must sell food and beverage products at the prices listed on the menu.

Specific variations from this general policy and any other significant variations must be approved by management.

### Proper Recordkeeping

The accurate recording of security-related information is essential for effective internal control. Asset security documents including purchase orders, inventory evaluation sheets, sales records, and payroll schedules should be designed to be easy to complete and understand.

### Written Policies and Procedures

Employees can only be expected to follow policies they understand. Significant security-related policies and procedures should be in writing and be included in employee manuals and new employee orientation programs for all employees. Written policy and procedure manuals also help ensure that all employees are treated fairly and similarly if policy violations occur.

### Physical Controls

Physical controls are often necessary to properly safeguard assets. Examples include security devices such as cash safes and locked storage areas.

### Performance Checks

Performance checks help to ensure all elements in the internal control system are functioning properly. Whenever possible, the checks should be independent (i.e., the person doing an internal verification should not be the same person responsible for collecting the data initially).

**Key Term**

**General ledger:** The main or primary accounting record of a business.

**Key Term**

**Cash receipts journal:** An accounting record used to summarize transactions related to cash receipts generated by a foodservice operation.

In large businesses, auditors who are independent of both operations and accounting personnel report directly to top-level management. In smaller operations, operators should conduct most, but not all, performance checks. For example, an operation's **bank reconciliation** should be done by personnel who are independent from those who initially account for cash receipts and vendor payments.

## Implementing and Monitoring a Revenue Security Program

In its simplest form, revenue control and security involve matching products sold with money received. Implementing and monitoring a total revenue security program involves ensuring that systems and recordkeeping are in place to allow foodservice operators to always verify that:

1) Documented Menu Item and Bar Orders = **Receivable Revenue**
2) Receivable Revenue = Total Guest Charges
3) Total Guest Charges = Sales Receipts
4) Sales Receipts = Sales (Bank) Deposits
5) Sales (Bank) Deposits = Funds Available to Pay Accounts Payable (AP)

**Key Term**

**bank reconciliation:** A process performed to ensure that a company's financial records are correct. This is done by comparing the company's internally recorded amounts with the amounts shown on its bank statement. Any differences must be justified. When there are no unexplained differences, accountants can affirm that the bank statement has been reconciled.

**Key Term**

**Receivable revenue:** The expected sales income when all food and beverage orders have been properly documented.

The term "receivable revenue" means that a specified amount (charges) must be included in Step 2 and actually paid for in Step 3.

To illustrate these five verification points required to ensure an effective revenue security program, consider Maoli Felize De la Cruz, who operates a Dominican Republic-themed restaurant in New York.

Maoli considers her restaurant to be a family-oriented establishment. It has a small (20 seat) cocktail area, and it seats 100 guests in the dining room. When she started the restaurant, she did not give much thought to the design of her revenue control systems because she generally worked in the operation. Due to her success, she now spends more time away from the property while developing a second restaurant, and she requires the security of an adequate revenue management system and the ability to review it quickly when she manages the first property.

Maoli has begun to develop a revenue security system by concentrating her efforts on ensuring that the five steps noted above are always followed:

Receivable Revenue = Total Guest Charges = Sales Receipts = Sales (Bank) Deposits = Funds Available to Pay Legitimate Accounts Payable (AP)

## Verification of Receivable Revenue

A key to verification of receivable revenue in a revenue security system is to follow one basic rule: "No menu item shall be issued from the kitchen or bar unless a permanent record of the item prepared has been made."

This means that the kitchen (or bar) should not fill any server request for menu items unless that product request has been documented in writing or electronically. In some small restaurants, the server's written request for food or beverages takes the form of a single (or multi-copy) written guest check that is designed specifically for the purpose of revenue control. The top copy of this multi-copy form generally is sent to the kitchen or bar. The guest check, in this case, becomes the documented request for the food or beverage product ordered.

This "paper-only" system can work, but it is subject to many forms of abuse and fraud. Therefore, when possible, foodservice operations should utilize a POS system in which the "guest check" consists of an electronic record of product requests and issues. Then a guest's order is viewed by the production staff on a computer screen or, in other cases, the POS system prints a hard copy of the order to be used by production staff filling the order.

In either case, the software within the POS system creates a permanent record of the transaction and issues a unique transaction number to identify the requested product(s). This record authorizes kitchen personnel to prepare food or the bartender to make a drink. If a foodservice operation elects to supply its employees with meals during work shifts, these meals should also be recorded in the POS system.

In the bar, the principle of verifying all receivable revenue is even more important. Bartenders should be instructed to never prepare a drink unless that drink has first been recorded in the POS system. This should be the standard operating procedure (SOP), even if the bartender is working alone. This rule regarding menu item ordering is important for two reasons. First, requiring a permanent documented order ensures that there is a record of each product's production. Second, this record can be used to verify both proper inventory usage and product sales totals.

Maoli enforces this basic rule by requiring that no item be served from her kitchen or the bar without the sale first being entered into her POS system. If her verification of product production works correctly, Maoli will note the following formula is always correct:

Documented Menu Item and Bar Orders = Receivable Revenue

---

**Technology at Work**

A high-quality POS system always provides detailed reports. The sales report dashboard on a POS system provides an overview of all transactions completed during a selected time period. This includes net sales, service charges, tips, total guests served, table turn times, and a breakdown of all service types and payment methods. It should also provide detailed sales exception reports that allow users to quickly see an overview of all voids, discounts, and refunds. The reports also allow users to identify the specific servers and operators who are giving and approving sales exceptions such as voiding receipts, discounting food, and offering refunds.

It is important to recognize that, regardless of its sophistication, a POS system will not "bring" control to a foodservice operation. A high-quality POS system can, however, take good control systems that have been carefully designed by management and add to them speed, accuracy, and/or additional information.

Properly utilized, a POS system is of immense value. If, however, an operation has no formal revenue security plan, the POS system simply becomes a high-tech adding machine used primarily to sum guest purchases and little more. Properly selected and utilized, however, POS systems play a crucial role in the implementation of an operation's complete revenue security system.

To review the cash control-related features of some currently popular POS systems, enter "POS revenue control reports for restaurants" in your favorite search engine and view the results.

---

Experienced foodservice operators know that, despite their best efforts, it is possible for employees to issue menu items without a documented product request when:

1) Two or more employees work together to defraud the operation. Collusion of this type can be discovered when operators use a system to carefully count the number of items removed from inventory and then compare that number to the number of products actually issued.
2) A single employee (such as a bartender working alone) is responsible for both making and filling the product request. If fraud occurs, operators can uncover it when they carefully compare the number of items (or, in some cases such as bottled beers and wine sold by the bottle, beverage servings) removed from inventory with the number of recorded product issues.

## Verification of Guest Charges

When the production staff is required to prepare and distribute products only in response to a properly documented request, it is critical that it (the documented request) results in charges to the guest. The reason: It makes little sense to enforce verification of product production without also requiring the service staff to ensure that guest charges match these requests.

When an operator insists that no products be issued without a POS-generated request, the managerial goal is to ensure that the menu items and/or beverage products equal guest check totals. In other words, all issued products should result in appropriate charges to the guest.

When properly implemented, this second step of the revenue control system will ensure the following formula always holds true:

$$\text{Receivable Revenues} = \text{Total Guest Charges}$$

Maoli has now implemented the first two key revenue control principles. The first one is that no items can be sold from the kitchen or bar unless the production order is documented, and the second one is that all guest charges must match revenue receivables generated by the menu items and/or beverage items served.

With these two systems in place, Maoli can deal with many problems. If, for example, a guest has "walked" the check, the operation has a duplicate record of the transaction. The POS will have recorded which products were sold to this guest, who sold them and, perhaps, additional information including time of the sale, number of guests in the party, and total sales value of the menu items produced.

The POS system Maoli uses also ensures that service personnel cannot "change" the prices charged for items sold. Note: This would likely be possible in an operation using manual (paper) guest checks.

To complete this aspect of her control system, Maoli implements a strict policy regarding the documentation of employee meals (a labor cost). Recall that this amount is needed to accurately compute cost of food sold (see Chapter 6) when the property's income statement is prepared.

Maoli is now ready to address the next major component in her revenue security system. That component is the actual collection of guest payments. The sum of these payments will represent Maoli's actual sales receipts.

## Verification of Sales Receipts

The term "sales receipts" refers to the actual revenue received by the cashier, server, bartender, or other designated personnel in payment for products served. In Maoli's case, this means all sales revenue from her restaurant and bar.

The essential principle to recognize in this step is that it requires two individuals (the cashier and a member of Maoli's management team) to verify sales receipts. Although this will not prevent possible collusion by a pair of individuals, it is important that sales receipt verification is a two-person process. Maoli wants to ensure that the amount of cash collected when added to her non-cash (including credit and debit card) guest payments matches the dollar amount she has charged her guests as recorded in the POS system.

In most operations, individual guest charges are recorded only in the POS system. This is the case, for example, in a QSR or cafeteria, where food purchases are totaled and paid for at the same time. In these instances, the POS system provides an accurate total of guest and other charges. Receipts collected should always equal these charges. If Maoli's revenue security system is working properly, the following formula noted above will always be true:

$$\text{Total Guest Charges} = \text{Sales Receipts}$$

Note that total POS-recorded charges will consist of all sales, service charges, tips, and guest-paid taxes that represent the total revenue the operation should receive.

## Verification of Sales Deposits

Most foodservice operations make a sales deposit each day the property is open because keeping excessive amounts of cash on hand is not advisable. It is strongly recommended that only management make the actual bank deposit of daily sales receipts. A cashier or other clerical assistant may complete a deposit slip, but the operator should be responsible for monitoring the deposit of sales.

This task involves the actual verification of the deposit's contents and the process of matching bank deposits with actual sales receipts. These two numbers should match. If Maoli or a member of her management team deposits Thursday's sales on Friday, the Friday deposit should match the sales amount of Thursday. If it does not, her operation has experienced some loss of revenue.

It is this step of the revenue control system in which embezzlement is most likely to occur. Embezzlement is a crime that often goes undetected for a long time because the embezzler is often a trusted employee. Falsification of, or destroying, bank deposits is a common method of embezzlement.

To prevent this activity, Maoli should take the following steps to protect her sales deposits:

✓ Make bank deposits of cash and checks daily, if possible.
✓ Ensure the person preparing and making the deposit is bonded.
✓ Establish written policies for completing bank reconciliations: the regularly scheduled comparison of the business's deposit records with the bank's deposit records.

✓ Payment card fund transfers to a business's bank account should be reconciled each time they occur. Today, in most cases, cash and non-cash payment reconciliations can be accomplished daily using online banking features.

✓ Review and approve written bank statement reconciliations at least monthly.

✓ Change combinations on office safes periodically and share the combinations with the fewest possible employees.

✓ Require all cash-handling employees to take regular and uninterrupted vacations at least annually so another employee can assume and potentially uncover any improper practices.

✓ Consider employing an outside auditor (see Chapter 4) to examine the accuracy of deposits on an annual basis.

If the verification of sales deposits is done correctly, and no embezzlement is occurring, the following formula should always hold true:

$$\text{Sales Receipts} = \text{Sales (Bank) Deposits}$$

Sales deposit records are maintained in an operation's **back office accounting system**, and this data must be regularly and carefully monitored by an operation's managers or owners.

## Key Term

**Back office accounting system:** The accounting software used to maintain a business's accounting records that are not contained in its POS system. This typically includes items such as payroll records, accounts payable and receivable, taxes due and payable, and net profit and loss summaries, as well as all balance sheet entries.

## Technology at Work

Modern POS systems are an important tool in maintaining revenue control, but they are not the only helpful item. To ensure an effective revenue control system, foodservice operators must also have an up-to-date and effective back office accounting system.

The term "back office" originated when early companies designed their offices so that the front portion was a workplace for the employees who interact with guests. The back portion of the office were the workplaces of associates with little guest interaction such as accounting clerks.

Back office accounting software is used by foodservice and other businesses to help track income and expenses, create invoices, calculate sales taxes due, price recipes and menus, and more.

The best back office accounting systems can be interfaced (electronically connected) to an operation's POS system. Doing so makes it easy to monitor both revenue collection and revenue deposits as well as cash account balances.

To learn more about the features included in these essential accounting systems, enter "back office accounting systems for restaurants" in your favorite search engine and review the results.

### Verification of Accounts Payable (AP)

AP (see Chapter 4) as defined in this step refers to the legitimate amount owed to a vendor for the purchase of products or services. The basic principle to be followed when verifying AP is:

"The individual authorizing the purchase should verify the legitimacy of the vendor's invoice before it is paid."

Vendor payments are often an overlooked potential threat to the security of a foodservice operation's revenue. Of course, an operation should pay all valid expenses. However, both external vendors and an operation's employees can attempt to defraud a foodservice operation by invoices.

For example, consider again the case of Maoli. She has just received an invoice for fluorescent light bulbs. The invoice is for over $400 dollars, but the invoice indicates that only two dozen bulbs were delivered, and this is a large overcharge.

Maoli is not familiar with this specific vendor, but the delivery slip included with the invoice was signed (six weeks ago!) by her receiving clerk. As a result, Maoli and her operation may be the victims of a vendor's invoice scam that threatens her operation's revenue.

---

**Find Out More**

Foodservice operations are popular targets for invoice fraud. In these scams, criminals send bills for goods or services a business never ordered or received. The scam succeeds mainly because the invoices look legitimate, and unsuspecting accounts payable employees do not look closely to determine if they are real. They simply make the payment, often thinking someone else in the operation placed the order.

To protect your operation from invoice scams, you must:

1) Be cautious when processing invoices: Ensure your accounts payable personnel are aware of these scams and are cautious when processing invoices.
2) Verify unfamiliar vendors and do not purchase from new suppliers or pay invoices from unfamiliar vendors until you verify their existence and reliability.
3) Check invoices against original purchase orders (POs) before paying them: a written PO (see Chapter 3) should exist for each invoice an operator receives.

To learn more about additional procedures and policies operators can implement to minimize the chances of becoming a victim of invoice fraud, enter "protecting against invoice fraud" in your favorite search engine and review the results.

---

1) No menu items shall be produced in the kitchen or bar unless a permanent record of the production is made.
2) Revenue receivables must equal total guest charges.
3) Both the cashier and the operator must verify all sales receipts.
4) Management must personally verify all bank deposits.
5) Management or the individual authorizing the purchase should verify that all vendor invoices represent accounts payable expenses.

**Figure 11.3**   Revenue Control Points Summary

Dishonest suppliers can take advantage of weaknesses in an organization's purchasing procedures and/or of unsuspecting employees who may not be aware of their fraudulent practices. In addition, the supplies delivered by these bogus firms are most often highly overpriced and of poor quality.

When her revenue security program is working properly, Maoli can confirm that:

Sales (Bank) Deposits = Funds Available to Pay Accounts Payable (AP)

Funds available for AP should only be used to pay legitimate expenses that result from a purchase verified by authorized personnel within the foodservice operation.

When Maoli properly completes the building of her revenue control system, its key features can be summarized as shown in Figure 11.3.

The best revenue control programs help foodservice operators maintain the security of their cash. Previous chapters of this book focused on controlling and optimizing product, service, and labor costs, and other operating expenses. Information in those chapters allows foodservice operators to better understand what these costs *are*.

The next, and final, chapter of this book addresses budgeting for and analyzing an operation's actual costs to help ensure these costs are consistently what they *should be*.

**What Would You Do? 11.2**

Steve Bart worked for 15 years as the head snack bar cashier for the Downtown Sports Arena Complex, a facility whose food concessions were managed by Rose Harper's "Elite Catering" company.

Steve had twice won the company's "Employee of the Year" award, and Rose considered Steve to be a valued and trusted employee who had, on many occasions, performed far above and beyond what was required.

Rose was very surprised when newly installed video surveillance equipment confirmed that Steve, despite strict written rules against it, had recently given free food and beverages to friends who visited the arena.

*(Continued)*

When confronted with the video evidence, Steve admitted the conduct, apologized profusely, and asked Rose for a second chance. He promised never to give free food or beverages to anyone in the future.

On the advice of the company's attorney, Rose is documenting, in writing, her decision on handling the situation.

**Assume you were Rose. Do you believe an employee caught defrauding their employer should ever be given a second chance? If so, under what circumstances? What impact will your decision in this case have if, in the future, other employees are caught stealing from your operation?**

## Key Terms

| | | |
|---|---|---|
| Fraud | Over (cash bank) | Bank reconciliation |
| Embezzlement | Bonding (employee) | Receivable revenue |
| Walk (bill) | Collusion | Back office accounting |
| Open check | General ledger | system |
| Short (cash bank) | Cash receipts journal | |

---

### Operator's 10-Point Tactics for Success Checklist

Evaluate your need for, and the current status of, each of the following operational tactics. For those tactics you think are important, but not yet in place, develop an action plan for its implementation including who will be responsible for the tactic's completion and the target date by which it should be completed.

| Tactic | Don't Agree (Not Done) | Agree (Done) | Agree (Not Done) | If Not Done — Who Is Responsible? | If Not Done — Target Completion Date |
|---|---|---|---|---|---|
| 1) Operator understands the importance to profits of developing an effective revenue control program. | —— | —— | —— | | |
| 2) Operator can identify external threats to revenue security in their own operation. | —— | —— | —— | | |

| | | | | If Not Done | |
| Tactic | Don't Agree (Not Done) | Agree (Done) | Agree (Not Done) | Who Is Responsible? | Target Completion Date |
|---|---|---|---|---|---|
| 3) Operator can identify internal threats to revenue security in their own operation. | ____ | ____ | ____ | | |
| 4) Operator can state the objectives of an internal revenue security program. | ____ | ____ | ____ | | |
| 5) Operator can summarize the elements required in an effective revenue control system. | ____ | ____ | ____ | | |
| 6) Operator recognizes the importance of continually monitoring and verifying menu and bar item issues. | ____ | ____ | ____ | | |
| 7) Operator recognizes the importance of continually monitoring and verifying guest charges. | ____ | ____ | ____ | | |
| 8) Operator recognizes the importance of continually monitoring and verifying sales receipts. | ____ | ____ | ____ | | |
| 9) Operator recognizes the importance of continually monitoring and verifying sales deposits. | ____ | ____ | ____ | | |
| 10) Operator recognizes the importance of continually monitoring and verifying the accounts payable (AP) system. | ____ | ____ | ____ | | |

# 12

# Budgeting and Analyzing Operating Results

**What You Will Learn**

1) The Importance of Operating Budgets
2) How to Create an Operating Budget
3) How to Compare Actual Operating Results to Budgeted Results

**Operator's Brief**

In this chapter, you will learn the importance of creating and properly monitoring your operating budget. The operating budget, or financial plan, is developed to help a business reach its future financial goals. Your operating budget will tell you what you must achieve to meet your predetermined cost and profit objectives.

To create an operating budget, you must first consider:

1) Prior-period operating results (if an existing operation)
2) Assumptions made about the next period's operation
3) The operation's financial objectives

An effective operating budget estimates your business's future sales, expenses, and profits for a specific accounting period. To estimate future sales most accurately, you must review historical records, and then consider any internal or external factors that may affect future revenue generation. The next step in the budgeting process is to estimate each fixed, variable, and mixed cost.

You will learn that one important result of budget creation is the ability to make comparisons between budgeted results and your operation's actual financial performance.

In this chapter, you will learn how to analyze actual financial performance and compare it to the results you originally planned (budgeted). You will also learn how to use this information to take corrective action(s) or to modify your initial operating budget. Doing so will better ensure that you and your operation will achieve your financial goals.

# The Importance of Operating Budgets

Just as the income statement tells operators about their past performance, the operating budget (see Chapter 1), or financial plan, is developed to help a business achieve its future financial goals. The operating budget tells foodservice operators what they must do to meet predetermined cost and profit objectives.

The utilization of an operating budget generally involves several activities:

1) Establishing realistic financial goals
2) Developing a budget (financial plan) to achieve goals
3) Comparing actual operating results with budgeted results
4) Taking corrective action, when necessary, to modify operational procedures and/or the financial plans

Preparing a budget and staying within its financial boundaries help operators meet their financial goals. Without this plan, operators must guess the amount to spend and how much sales should be anticipated. Effective operators build their budgets, monitor them closely, and modify them when necessary to achieve desired results.

## Types of Operating Budgets

One helpful way to consider the purpose of an operating budget is by its coverage (time frame). While an operating budget may be prepared for any accounting period desired, operating budget lengths are typically considered to be one of the three types shown in Figure 12.1.

Regardless of their time frames, all operating budgets are developed by planning estimated revenues, expenses, and profits associated with operating a foodservice business.

| Type of Operating Budget | Budget Characteristics |
|---|---|
| Long-range | Typically prepared for a period of up to five years. While not highly detailed, it provides financial views and financial projections related to long-term goals. |
| Annual | Typically prepared for one calendar or fiscal year or, in some cases, one season. The budget may consist of 12 months or 13 periods of 28 days each. |
| Achievement | Prepared for a limited time, often a month, a week, or even one day. It most often provides very current operating information and greatly assists in making current operating decisions. |

**Figure 12.1**   Operating Budgets

## Advantages of Operating Budgets

The owners of a foodservice operation want to know what they should expect to earn from investments, and a budget helps to project those earnings. Questions about the amount of revenue likely generated, the amount of cash that should be available for bill payment or distribution as earnings, and the proper timing of major purchases can all be addressed in a properly developed budget. As a result, in organizations of all sizes, proper budgeting is an essential process that must be carefully planned and implemented.

The advantages of preparing and using an operating budget are many and are summarized in Figure 12.2.

1) It is a way to analyze alternative courses of action and allows operators to examine alternatives before adopting a specific plan.
2) It requires operators to examine the facts about what should be their desired profit levels.
3) It enables operators to define standards used to develop and enforce appropriate cost control systems.
4) It allows operators to anticipate and prepare for future business conditions.
5) It helps operators periodically carry out a self-evaluation of their business and its progress toward meeting financial objectives.
6) It is a communication channel that allows the operation's objectives to be passed along to stakeholders including owners, investors, managers, and staff.
7) It encourages those who participated in budget preparation to establish their own operating objectives, evaluation tactics, and tools.
8) It provides operators with reasonable estimates of future expense levels and serves an important aid when appropriate selling prices are determined.
9) It identifies time periods in which operational cash flows may need to be supplemented.
10) It communicates realistic financial performance expectations of operators to the owners of and investors in the business.

**Figure 12.2**   Advantages of Preparing and Using an Operating Budget

One way to consider operating budgets is to compare them to an operation's income statement (see Chapter 1). Recall that the income statement details the actual revenue, expenses, and profits incurred in operating a business. The operating budget is an operator's best estimate of all (or any portion of) a future income statement.

---

**Find Out More**

Some foodservice operators responsible for budget preparation for one or more units find membership in the Hospitality Financial and Technology Professionals (HFTP®) to be helpful.

Established in 1952, HFTP is an international, non-profit association head-quartered in Austin, Texas, with offices in the United Kingdom, Netherlands, and Dubai. HFTP is recognized as the professional group representing the finance and technology segments of the hospitality industry. HFTP offers members continuing education courses including those addressing budgeting through its HFTP Academy.

To find out more about the educational resources offered by this professional group and its "Certified Hospitality Accountant Professional (CHAE)" program, enter "HFTP Academy" in your favorite search engine and view the results.

---

## Creating an Operating Budget

Some operators believe it is difficult to develop an operating budget, and they do not take the time to do so. Creating an operating budget, however, does not need to be a complex process.

Before operators can begin to develop a budget, they must understand the essentials required for its creation. If these are not addressed before the operating budget is developed, the budgeting process that follows is not likely to yield an accurate or helpful financial planning tool.

Before beginning creating an operating budget, its developers must have available and understand the following:

1) Prior-period operating results (if an existing operation)
2) Assumptions about the next period's operations
3) Knowledge of the organization's financial objectives

1) Prior-Period Operating Results

The task of budgeting is easier when an operator knows the results from prior accounting periods. Experienced foodservice operators know that what occurred in their units in the past often indicates what may occur in the future. The further back and the greater detail with which an operator can track historical revenues and expenses, the more accurate a budget being planned will be.

For example, if operators know the revenues and expenses for the past 50 Saturdays, they are better able to forecast this coming Saturday's revenue and expense budgets than if they have only the last two Saturdays' data available.

When preparing a budget, **historical data** should always be considered along with the most recent data available.

For example, assume a foodservice operator knows that their revenues have, on average, increased 5% each month from the same period last year. However, in the last two months, the increase has been closer to zero. This may mean that the revenue increase trend has slowed or even stopped completely. Good operators modify historical trends by closely examining current conditions. In this example, the operator should probably estimate next month's revenue increases to be closer to 0% than to 5%. Note: While historical data is useful for ongoing operations, such data will not be available for new operations.

| Key Term |
| --- |
| **Historical data:** Information about an operation's past financial performance including revenue generated, expenses incurred, and profits (or losses) realized. |

2) Assumptions about the Next Period's Operations

Evaluating future conditions and business activity is always a key part of operating budget development. Examples include the opening of new competitive restaurants in the immediate area, scheduled occurrences including local sporting events and concerts, and significant changes in operating hours. Local newspapers, trade or business associations, and Chambers of Commerce are possible sources of helpful information about changes in future demand.

When significant changes are planned for an operation such as the introduction of new menu items, changes in operating hours, or the estimated impact of significant marketing efforts, assumptions about the impact of these actions become important. After these factors have been considered, realistic assumptions about changes in potential revenues and expenses may be made.

3) Knowledge of the Organization's Financial Objectives

An operation's financial objectives may consist of a specific profit target defined as a percent of revenue or a total dollar amount and specific financial and operational ratios (see Chapter 8) to be achieved. Some financial objectives are determined by an operation's owner(s) based on desired return on investment (ROI), and the operating budget must also address these goals.

The operating budget is actually a detailed plan that can be expressed by the basic formula:

Budgeted Revenue – Budgeted Expense = Budgeted Profit

The budgeted profit level an operator desires can be achieved when the operation realizes its budgeted revenue levels and spends only what is budgeted to generate the sales. If revenues fall short of forecast and/or if expenses are not reduced to match the shortfall, budgeted profit levels are not likely to be achieved.

In a similar manner, if actual revenues exceed forecasted levels, some expenses will also increase. If the increases are monitored carefully and are not excessive, increased profits should result. If, however, an operator allows actual expenses to exceed the levels required by the additional revenue, budgeted profits may not be achieved.

To illustrate the operating budget development process, consider Randy, the owner/operator of Randy's Restaurant. He is developing the operating budget for next year and has determined historical budget data essentials as shown below:

1) He has gathered his prior year operating results.
2) From information applicable to the area's economic conditions and his competitors, he has made the following assumptions about next year's operations:
   - Total revenues received will increase by 4% primarily because of a 4% menu price increase.
   - Food and beverage costs will increase by 3% due to inflation affecting food and beverage product prices. As a result, his product expense (cost of sales) targets are a 35% food cost and a 16% beverage cost, and a combined weighted food and beverage total cost of sales of 31.2% (food cost/food sales + beverage cost/beverage sales = 31.2%)
   - Management, staff wages, and benefits costs have a target of 35% of sales for total labor cost.
   - Prime cost will be 66.2% of sales.
   - All other controllable expenses will total no more than 12.4% of sales.
3) Randy's financial objectives for the restaurant are to earn profits (net income before income taxes) of at least 11.1% of sales, and net income (after taxes) of 8.3% of sales.

Given these assumptions, Randy is now ready to create the 12-monthly revenue and expense forecasts needed to complete next year's operating budget.

## Revenue Forecasts

Accurately forecasting an operation's revenues is critical because all forecasted expenses and profits will be based on revenue forecasts. In most cases, revenues should be estimated on a monthly (or weekly) basis. Then they can be combined to create the annual revenue budget. The reason: Many hospitality operations have seasonal revenue variations.

For example, a restaurant doing a significant amount of business in its outside dining patio area may generate reduced sales during times of the year when inclement weather makes outside dining less desirable for guests. Similarly, a restaurant operated in a ski resort town will generally be busier in the winter season than in summer months.

Forecasting revenues is not an exact science. However, it can be made more accurate when operators:

✓ **Review historical records**. Operators begin the revenue forecasting process by reviewing their revenue records from previous years. When an operation has been open at least as long as the budget period being developed, its revenue history can be extremely helpful in predicting future revenue levels.

✓ **Consider internal factors affecting revenues**. In this step, operators must consider any significant changes in the type, quantity, and direction of their marketing efforts. Other internal activities that can impact future revenues include those related to facility renovation that might affect dining capacity. In some operations, the number of hours to be open and/or menu prices may change. Any internal change an operator believes will likely impact future revenues should be considered in this step.

✓ **Consider external factors affecting revenues**. There are numerous external issues that could affect an operation's revenue forecasts. These include planned competitors' openings or closings, and other factors such as road improvements or construction that disrupts normal traffic patterns. Other factors that may yield revenue forecast changes are forecasted economic upturns or downturns that affect how potential guests may spend their **discretionary income** on dining services.

Returning to the example of Randy's Restaurant and using September as an example of the month he is now working on, Randy has reviewed last year's data for the month of September and found that his sales that month were $192,308, with an average guest check of $12.02.

---

**Key Term**

**Discretionary income:** Money a household or individual has available to invest, save, or spend after necessities are paid.

Examples of necessities include the cost of housing, food, clothing, utilities, and transportation.

He considers his internal and external factors affecting revenues and has estimated a net 4% increase in revenues for the coming September. Randy then computes his revenue forecast for the upcoming September as:

Sales Last Year × (1.00 + % Increase Estimate) = Revenue Forecast

Or

$192,308 × (1.00 + 0.04) = $200,000

Using his historical data, Randy knows that approximately 80% of his sales are from food, and 20% of his sales are from beverages. Therefore, he estimates $160,000 ($200,000 × 0.80 = $160,000) for food sales and $40,000 ($200,000 × 0.20 = $40,000) for beverage sales.

Randy's average sale per guest (see Chapter 2) for food and beverages for last September was $12.02. With a forecasted increase of 4% in selling prices, the forecast for this September's average sale per guest (check average) is calculated as:

Guest Check Average Last Year × (1 + % Increase Estimate) = Guest Check Average Forecast

Or

$12.02 × (1.00 + 0.04) = $12.50

With forecasted sales of $200,000 and a forecasted guest check average of $12.50, Randy's forecasted number of guests would be 16,000 ($200,000 revenue ÷ $12.50 guest check average = 16,000 guests).

In most cases, it is not realistic to assume an operator can forecast their business's exact monthly revenue one year in advance. With practice, accurate historical sales data, and a realistic view of internal and external variables that can impact an operation's future revenue generation, many operators can attain operating budget forecasts that are easily and routinely within 5–10% (plus or minus) of actual results for a forecasted accounting period.

---

### Technology at Work

Today's readily available restaurant sales forecasting tools help foodservice operators and owners better estimate future volume and make more informed and accurate decisions about purchasing and staffing.

These advanced technology tools, often interfaced with an operation's point-of-sale (POS) system, can use historical data to help foresee upcoming seasonality trends and predict how the trends will likely affect an operation's revenue generation. Since they assist in establishing realistic sales and profit goals, the result is better financial planning and preparation for the future.

To examine information about some systems that selected companies offer to assist in foodservice sales forecasting, enter "restaurant revenue forecasting software" in your favorite search engine and review the results.

## Expense Forecasts

Operators must budget for each fixed, variable, and **mixed cost** when they address individual expense categories on income statements. Fixed costs are simple to forecast because items such as rent, depreciation, and interest typically do not vary from month to month.

Variable costs, however, are directly related to the amount of revenue produced by a foodservice operation. For example, an operation that forecasts sales of 100 prime rib dinners on Friday night will have higher food and server (labor) costs than the operator in a similar facility that forecasts the sale of only 50 prime rib dinners. The reason: Variable expenses such as food, beverage, and labor costs are affected by sales levels.

Mixed costs, however, contain both fixed and variable cost components. For example, an operation will have a minimum number of employees (fixed costs) who must be scheduled to work even when business is slow. However, as the number of guests served increases, the operation will spend additional labor dollars (variable costs) to properly serve its guests. In this example, labor would be a mixed cost.

### Key Term

**Mixed cost:** A cost composed of a mixture of fixed and variable components. Costs are fixed for a set level of sales volume or activity and then become variable after this level is exceeded. Also referred to as a semi-fixed or semi-variable cost.

### Forecasting Fixed Costs

In most cases, budgeting fixed costs is easy. For example, Randy knows his monthly rent (occupancy) payments are $10,000. Creating the annual budget for "rent" expense is a simple matter of multiplication: $10,000 a month for 12 months yields a total annual occupancy cost of $120,000 ($10,000 × 12 months = $120,000). In this situation, since rent is a fixed cost, it remains unchanged regardless of the revenues generated by the restaurant. Note: Any anticipated increases in fixed costs for the coming year must be budgeted separately for each month of the year.

### Forecasting Variable Costs

Variable costs increase or decrease as revenue volumes change. For example, consider the cost of linen napkins used at Randy's Restaurant. As more guests are served, more napkins will be used, and higher napkin expenses will be incurred because of laundry charges. The laundry charges for napkins in this situation represents variable costs. The food Randy will purchase, the beverages he will serve, and a variety of other expenses are all variable costs that increase as the number of guests served increases.

Variable costs can be forecasted using targeted percentages or costs per unit (guests served). For example, food costs might be forecasted at a targeted food cost percentage. Similarly, an operator might forecast cleaning supplies as a percentage of total sales. When percentages are used, the sales forecast is multiplied by the target cost percentage to arrive at the forecasted cost.

In the case of Randy's Restaurant, a targeted food cost percentage of 35% and $160,000 in estimated food sales would yield the following forecasted cost of sales for food:

Revenue Forecast × Targeted Food Cost % = Forecasted Cost of Sales: Food

Or

$$\$160,000 \times 0.35 = \$56,000$$

Randy will use this same approach and formula to estimate beverage costs and all other variable expenses.

**Forecasting Mixed Costs**

To create the expense portion of his operating budget, Randy must accurately forecast his mixed costs. One of the largest line-item costs in many foodservice operations is that of labor. It is a mixed cost because it includes hourly wages (variable costs), salaries (fixed costs), and **employee benefits** (mixed costs).

Experienced operators know that, when labor costs are excessive, profits are reduced. As a result, this is an area of budgeting and cost control that is extremely important. Accurate budgets used to help control future labor costs can be precisely calculated using a three-step method.

> **Key Term**
>
> **Employee benefits:** Any form of rewards or compensation provided to employees in addition to their base salaries and wages. Benefits offered to employees may be legally mandated or voluntary.

Step 1: *Determine Targeted Total Labor Dollars to Be Spent.*

In most cases, the determination of total labor costs is tied to the targeted or standard costs an operation desires. These standards or goals may be established by considering the historical performance of an operation, by referring to industry segments or company averages, or by considering the profit level targets of the business. In most cases, the standard will be developed by considering each of these important factors.

When all relevant labor-related information has been considered, a labor cost percentage standard (see Chapter 8) is used to determine the operating budget for total labor.

Randy has set his total labor cost standard to be 35% of total sales. With a $200,000 sales forecast for September, and a 35% labor cost percentage standard, the amount to be budgeted for total labor (salaries, wages, and employee benefits) for September is calculated as:

$$\text{Revenue Forecast} \times \text{Labor Cost \% Standard} = \text{Forecasted Total Labor Cost}$$

Or

$$\$200,000 \times 0.35 = \$70,000$$

Step 2: *Subtract the Cost of Employee Benefits.*

Employee benefits consist of costs associated with, or allocated to, payroll, and they must be paid by employers. Examples include items such as payroll taxes (e.g., mandated contributions to Social Security and worker's compensation plans) and voluntary benefit programs offered by the operation.

The costs of these mandatory and voluntary benefit programs can be significant, and they can include costs such as those for:

✓ Performance bonuses
✓ Health, dental, and vision insurance
✓ Life insurance
✓ Long-term disability insurance
✓ Employee meals
✓ Sick leave
✓ Paid holidays
✓ Paid vacation

Foodservice operators may offer all or some of these benefits; however, the applicable mandatory and voluntary payroll allocations must be subtracted from the labor dollars available to be spent when developing an accurate operating budget.

For Randy's Restaurant, assume that, historically, mandatory and voluntary benefits have accounted for 20% of the total labor costs incurred by the operation. The calculation required to determine the budgeted payroll allocation (employee benefits) amount would then be:

$$\text{Forecasted Labor Cost} \times \text{Employee Benefits \%} = \text{Budgeted Employee Benefits}$$

Or

$$\$70,000 \times 0.20 = \$14,000$$

In this example, the amount remaining for use in paying all operational salaries and hourly wages (budgeted payroll) would be computed as:

$$\text{Forecasted Labor Cost} - \text{Budgeted Employee Benefits} = \text{Budgeted Payroll}$$

Or

$$\$70,000 - \$14,000 = \$56,000$$

Step 3: *Subtract Management (Fixed) Costs to Determine Staff (Variable) Costs.*
In most cases, fixed payroll remains unchanged from one pay period to the next pay period unless an individual receiving fixed pay separates employment from the organization or is given a raise. Management labor costs are typically a fixed cost, and operators are paid a fixed salary. Staff costs, alternatively, consist primarily of those dollars paid to hourly employees. Variable payroll is the amount that "varies" with changes in sales volume because these employees are paid a wage (hourly rate) based on the number of hours worked during a specific pay period.

The distinction between fixed and variable labor is an important one since managers may sometimes have little control over their fixed labor costs while exerting nearly 100% control over variable labor costs.

To determine the amount of money to be budgeted for (hourly) staff, Randy must first subtract the management (fixed) portion of his labor costs. This fixed labor component consists of all the operational salaries he will pay. These labor costs must be budgeted and subtracted from the total available for labor because the salary amounts will be paid regardless of sales volume.

To illustrate the budgeting process, assume Randy pays $18,000 in salaries each month. The amount to be budgeted for staff labor costs for September would be calculated as:

Budgeted Payroll − Management Costs = Budgeted Staff Costs

Or

$56,000 − $18,000 = $38,000

As shown above, $38,000 will be available to be paid to Randy's front-of-house and back-of-house hourly paid employees.

In most foodservice operations, those who successfully create an operating budget for their food, beverages, and labor expenses will have accounted for more than 50% of their total costs. All other expenses can be budgeted using the same methods for fixed, variable, and mixed costs.

By successfully forecasting his revenue and expenses, Randy can now develop the entire budget for his operation for September. Recall that Randy's assumptions and financial objectives were:

✓ Total revenues received will increase by 4% primarily due to a 4% menu price increase.
✓ Food and beverage costs will increase by 3%. Targets are a 35% food cost and a 16% beverage cost, for a total combined food and beverage cost of sales of 31.2%.
✓ Management, staff wages, and benefits have a target of 35% of sales for total labor cost.
✓ All other controllable expenses will total no more than 12.4% of sales.
✓ Net income (after taxes) will be 8.3% of sales.

Based on this information, the resulting budget for Randy's Restaurant for September of the budget year is shown in Figure 12.3.

**Randy's Restaurant**
Budget for September
Budgeted Number of Guests = 16,000

| | Next Year | % |
|---|---|---|
| **SALES** | | |
| Food | $160,000 | 80.0 |
| Beverage | $ 40,000 | 20.0 |
| Total Sales | $200,000 | 100.0 |
| **COST OF SALES** | | |
| Food | $ 56,000 | 35.0 |
| Beverages | $ 6,400 | 16.0 |
| Total Cost of Sales | $ 62,400 | 31.2 |
| **LABOR** | | |
| Management | $ 18,000 | 9.0 |
| Staff | $ 38,000 | 19.0 |
| Employee Benefits | $ 14,000 | 7.0 |
| Total Labor | $ 70,000 | 35.0 |
| **PRIME COST** | $132,400 | 66.2 |
| **OTHER CONTROLLABLE EXPENSES** | | |
| Direct Operating Expenses | $ 7,856 | 3.9 |
| Music & Entertainment | $ 1,070 | 0.5 |
| Marketing | $ 3,212 | 1.6 |
| Utilities | $ 5,277 | 2.6 |
| General & Administrative Expenses | $ 5,570 | 2.8 |
| Repairs & Maintenance | $ 1,810 | 0.9 |
| Total Other Controllable Expenses | $ 24,795 | 12.4 |
| **CONTROLLABLE INCOME** | $ 42,805 | 21.4 |
| **NON-CONTROLLABLE EXPENSES** | | |
| Occupancy Costs | $ 10,000 | 5.0 |
| Equipment Leases | $ — | 0.0 |
| Depreciation & Amortization | $ 3,400 | 1.7 |
| Total Non-Controllable Expenses | $ 13,400 | 6.7 |
| **RESTAURANT OPERATING INCOME** | $ 29,405 | 14.7 |
| Interest Expense | $ 7,200 | 3.6 |
| **INCOME BEFORE INCOME TAXES** | $ 22,205 | 11.1 |
| Income Taxes | $ 5,551 | 2.8 |
| **NET INCOME** | $ 16,654 | 8.3 |

**Figure 12.3** Randy's Restaurant Operating Budget for September Next Year

As can be seen in Figure 12.3, every fixed, variable, and mixed cost, considered individually, must be included in the operating budget. When properly completed, the result will be an operating budget that:

1) Is based upon a realistic revenue estimate
2) Considers all known fixed, variable, and mixed costs
3) Is intended to achieve the organization's financial goals
4) Can be monitored to ensure adherence to the budget's guidelines
5) May be modified, when necessary

Annual operating budgets are most often the compilation of monthly operating budgets. Therefore, operators may use their monthly operating budgets to create annual budgets and monitor weekly (or even daily) versions of their overall operating budgets.

---

**What Would You Do? 12.1**

"I just don't' know," said Trishauna, "it might make an impact, but it might not make a very big one. I think we'll just have to wait and see."

Trishauna was talking to Jill, her partner in the Jacked Up Coffee Bar, a coffee and pastry shop located adjacent to the State University campus.

Trishauna and Jill were preparing next year's operating budget. They had just begun the process and were forecasting their next year's revenues. Both Trishauna and Jill were aware that a major coffee shop chain had just announced plans to open a new shop within one block of the Jacked Up Coffee Bar.

"Well," said Jill, "I think our clientele is loyal. I know our new competitor will generate some business, but I'm not sure how much of that business will come from us versus other coffee shops in the area."

"Exactly," replied Trishauna. "That's why I think it'll be a real challenge for us to forecast our revenue for next year."

**Assume you were the owners of the Jacked Up Coffee Bar. How important will it be for you to consider external influences such as the opening of a new competitor as you create your next year's revenue budget? What could be a likely result if you did not make such considerations?**

---

## Monitoring the Operating Budget

An operating budget details a plan for future financial activities. In many cases, an operator's forecast of future results will be reasonably accurate, and, in other cases, it will not be as accurate. For example, revenue may not reach forecasted levels, expenses may exceed estimates, and internal or external factors not considered when the operating budget was prepared may negatively or positively impact financial performance.

An operation's budget will have little value if management does not utilize it. In general, the operating budget should be regularly monitored in each key area:

✓ Revenue
✓ Expenses
✓ Profit

### Revenue

If sales should fall below projected levels, the impact on profit can be substantial, and it may be impossible to meet profit goals. If revenues consistently exceed projections, variable cost portions of the budget must be modified or, ultimately, these expenses will soon exceed their budgeted dollar amounts. Effective operators compare their actual sales to those they have projected on a regular basis.

An appropriate revenue comparison must include a comparison of both total sales and guest check averages. The reason: An operation's sales may increase at the same time its guest count is increasing, or an operation's sales may increase with a declining guest count (if the average sale per guest increases significantly). While both situations will result in an increase in total revenue, it is important for foodservice operators to understand the source of their revenue increases.

### Expenses

Foodservice operators must be careful to monitor their operating expenses because costs that are too high or too low may be cause for concern. Just as it is not possible to estimate future sales volumes perfectly, it is also not possible to estimate future expenses perfectly, especially since some expenses will vary as sales volumes increase or decrease.

As business conditions change, revisions in the operating budget are to be expected. This is true because operating budgets are based on a specific set of assumptions and, as the assumptions change, so too will the accuracy of the operating budget produced from those assumptions.

To illustrate, assume an operator budgeted $1,000 in January for snow removal from the parking lot attached to a restaurant operating in upper New York State. If unusually severe weather causes the operator to spend $2,000 for snow removal in January, the assumption (normal levels of snowfall) was incorrect, and the original budget will be incorrect as well.

*Profit*

An operation's forecasted profits must be realized if the operation is to provide adequate returns for its owner(s). To illustrate, consider the case of Laura, the manager of a foodservice establishment with excellent sales but below-budgeted profits. For this year, Laura budgeted a 5% profit on $2,000,000 of sales; and therefore, a $100,000 profit ($2,000,000 × 0.05 = $100,000) was anticipated.

At year's end, Laura achieved her sales goal, but in doing so she generated only $50,000 profit, or 2.5% of sales ($50,000 ÷ $2,000,000 = 2.5%). If this operation's owners feel that $50,000 is an adequate return for their investment and risk, Laura's services may be retained. If they do not, she may lose her position, even though she operates a "profitable" restaurant. The reason: Management's task is not merely to generate a profit, but rather to generate the realistic profit level that was planned.

## Comparing Planned Results to Actual Results

One important task of a foodservice operator is to optimize profitability by analyzing the differences between planned for (budgeted) results and actual operating results. To do this effectively, operators must receive timely income statements that accurately detail the actual operating results. When they do, these actual results can then be compared to the operating budget for the same accounting period. The difference between planned (budgeted) results and actual results is called budget **variance**.

The basic formula operators use to calculate a budget variance is:

Actual Results − Budgeted Results = Variance

A variance may be expressed in either dollar or percentage terms, and it can be either positive (favorable) or negative (unfavorable). A **favorable variance** occurs when the variance is an improvement on the budget (revenues are higher or expenses are lower). An **unfavorable variance** occurs when actual results do not meet budget expectations (revenues are lower or expenses are higher).

**Key Term**

**Variance:** The difference between an operation's actual performance and its planned (budgeted) performance.

**Key Term**

**Favorable variance:** Better-than-expected performance when actual results are compared to budgeted results.

**Key Term**

**Unfavorable variance:** Worse-than-expected performance when actual results are compared to budgeted results.

For example, if the budget for carpet cleaning is $1,000 for a given month, but the actual expenditure for those services is $1,250, the variance is calculated as:

Actual Results − Budgeted Results = Variance

Or

$1,250 − $1,000 = $250

In this example, the variance may be expressed as a dollar amount ($250) or as a percentage of the original budget. The computation for a percentage variance is:

$$\frac{\text{Variance}}{\text{Budgeted Results}} = \text{Percentage Variance}$$

Or

$$\frac{\$250}{\$1,000} = 0.25\,(25\%)$$

In this example, the variance is unfavorable to the operation because the actual expense is higher than the budgeted expense. In business, a variance can also be considered as either significant or insignificant. It is the operator's task to identify and address significant variances between budgeted and actual operating results.

A **significant variance** may be defined several ways; however, a common definition is that a significant variance is any difference in dollars or percentage between budgeted and actual operating results that warrants further investigation. Significant variance is an important concept because not all variances need to be investigated.

**Key Term**

**Significant variance:** A variance that requires immediate management attention.

For example, assume that, at the beginning of a year, a foodservice operator prepares an annual (12-month) operating budget that forecasts the operation's December utility bill will be $6,000. When, 12 months later, the December bill arrives, it totals $6,420, and therefore it is $420 over budget ($6,420 actual expense − $6,000 budgeted expense = $420 variance).

Given the amount of the bill ($6,420) and the difficulty of accurately estimating utility expenses one year in advance, a difference of only $420 or 7% ($420 ÷ $6,000 = 0.07 or 7%) probably does *not* represent a significant variance from the operation's budget.

Alternatively, assume that the same operation had estimated office supplies usage at $100 for that same month, but the actual cost of supplies was $520. Again, the difference between the budgeted expense and actual expense is $420 ($520 actual − $100 budgeted = $420 variance). The office supplies variance, however, represents a very significant difference of 420% ($420 ÷ $100 = 4.20 or 420%) between planned and actual results, and it probably should be thoroughly investigated.

Foodservice operators must determine what represents a significant variance based on knowledge of their specific operations along with applicable company policies and procedures. Small percentage differences can be important if they represent large dollar amounts. Similarly, small dollar amounts can be significant if they represent large percentage differences from planned results. Variations from budgeted results can occur in revenues, expenses, and profits. Operators can monitor all these areas using a four-step operating budget monitoring process.

Step 1   Compare actual results to the operating budget.
Step 2   Identify significant budget variances.
Step 3   Determine causes of significant budget variances.
Step 4   Take corrective action or modify the operating budget.

In Step 1, an operator reviews the income statement and operating budget data for a specified accounting period. In Steps 2 and 3, actual operating results are compared to the budget and significant variances, if any, are identified and analyzed. Finally, in Step 4, corrective action is taken to reduce or eliminate unfavorable variances, or if it is appropriate to do so, the budget is modified to reflect the new realities confronting the business.

In most cases, operators should compare their actual results to their operating budget results in each of the income statement's three major sections of sales, expense, and profits. To illustrate the process, consider again the operating budget for Randy's Restaurant. Figure 12.4 shows Randy's original operating budget (see Figure 12.3) and his actual operating results in dollars and percentages of sales for the month of September.

If Randy properly monitors his budget, sales is the first area he will examine when comparing his actual results to his budgeted results. This is true because, if sales fall significantly below projected levels, there will likely be a significant negative impact on profit goals.

Secondly, when sales vary from projections, variable costs also fluctuate. In cases where sales are lower than budget projections, variable costs should most often be *less* than budgeted. In addition, when actual sales fall short of budgeted levels, fixed and mixed expenses such as rent and labor incurred by the operation will represent a larger-than-originally budgeted *percentage* of total sales. Alternatively, when actual sales exceed the budget, the total dollar value of variable expenses will increase, and the fixed and mixed expenses incurred should, if properly managed, represent a smaller-than-budgeted percentage of total revenue.

A close examination of Figure 12.4 shows that Randy has experienced a shortfall in both food and beverage revenue when compared to his operating budget.

Randy's Restaurant
Budget versus Actual Comparison for September

| | Budgeted Number of Guests = 16,000 | | Actual Number of Guests = 15,500 | |
| --- | --- | --- | --- | --- |
| | Budget | % | Actual | % |
| **SALES** | | | | |
| Food | $160,000 | 80.0% | $150,750 | 79.3% |
| Beverage | $ 40,000 | 20.0% | $ 39,250 | 20.7% |
| Total Sales | $200,000 | 100.0% | $190,000 | 100% |
| **COST OF SALES** | | | | |
| Food | $ 56,000 | 35.0% | $ 53,800 | 35.7% |
| Beverages | $ 6,400 | 16.0% | $ 6,300 | 16.1% |
| Total Cost of Sales | $ 62,400 | 31.2% | $ 60,100 | 31.6% |
| **LABOR** | | | | |
| Management | $ 18,000 | 9.0% | $ 18,000 | 9.5% |
| Staff | $ 38,000 | 19.0% | $ 37,800 | 19.9% |
| Employee Benefits | $ 14,000 | 7.0% | $ 11,160 | 5.9% |
| Total Labor | $ 70,000 | 35.0% | $ 66,960 | 35.2% |
| **PRIME COST** | $132,400 | 66.2% | $127,060 | 66.9% |
| **OTHER CONTROLLABLE EXPENSES** | | | | |
| Direct Operating Expenses | $ 7,856 | 3.9% | $ 7,750 | 4.1% |
| Music & Entertainment | $ 1,070 | 0.5% | $ 1,070 | 0.6% |
| Marketing | $ 3,212 | 1.6% | $ 1,350 | 0.7% |
| Utilities | $ 5,277 | 2.6% | $ 5,195 | 2.7% |
| General & Administrative Expenses | $ 5,570 | 2.8% | $ 5,455 | 2.9% |
| Repairs & Maintenance | $ 1,810 | 0.9% | $ 1,925 | 1.0% |
| Total Other Controllable Expenses | $ 24,795 | 12.4% | $ 22,745 | 12.0% |
| **CONTROLLABLE INCOME** | $ 42,805 | 21.4% | $ 40,195 | 21.2% |
| **NON-CONTROLLABLE EXPENSES** | | | | |
| Occupancy Costs | $ 10,000 | 5.0% | $ 10,000 | 5.3% |
| Equipment Leases | $ — | 0.0% | $ — | 0.0% |
| Depreciation & Amortization | $ 3,400 | 1.7% | $ 3,400 | 1.8% |
| Total Non-Controllable Expenses | $ 13,400 | 6.7% | $ 13,400 | 7.1% |
| **RESTAURANT OPERATING INCOME** | $ 29,405 | 14.7% | $ 26,795 | 14.1% |
| Interest Expense | $ 7,200 | 3.6% | $ 7,200 | 3.8% |
| **INCOME BEFORE INCOME TAXES** | $ 22,205 | 11.1% | $ 19,595 | 10.3% |
| Income Taxes | $ 5,551 | 2.8% | $ 4,899 | 2.6% |
| **NET INCOME** | $ 16,654 | 8.3% | $ 14,696 | 7.7% |

**Figure 12.4** Randy's Restaurant Budget versus Actual Comparison for September

One revenue problem Randy faced in September is that he budgeted for 16,000 guests and only served 15,500 guests.

Also, Randy budgeted for a $12.50 average guest check ($200,000 budgeted sales/16,000 budgeted guests = $12.50 budgeted guest check average), but he achieved a guest check average of only $12.26 ($190,000 actual sales/15,500 actual guests = $12.26 actual guest check average).

If sales consistently fall short of forecasts, Randy must evaluate all aspects of his entire operation to identify and correct the revenue shortfalls. Foodservice operations that consistently fall short of their sales projections must also evaluate the validity of the primary assumptions used to produce the sales portion of their operating budgets.

---

### Find Out More

To achieve desired sales levels, foodservice operators must have an effective marketing plan in place. Increasingly, an operation's marketing efforts must include both traditional approaches and newer Internet-based approaches.

While there are many publications addressing general marketing strategies for businesses, few publications exclusively address the marketing needs of foodservice operations.

One of the best and most up-to-date marketing resources available to foodservice operators, and one that exclusively addresses the on-site and online marketing of foodservice operations is *Marketing in Foodservice Operations* (January 2024) by David K. Hayes and Jack D. Ninemeier.

To learn more about the content and availability of this extremely valuable publication, enter "Wiley Marketing in Foodservice Operations" in your favorite search engine and review the results.

---

### Expense Analysis

Identifying significant variances in expenses is perhaps the most critical part of the budget monitoring process because many types of operating expenses are controllable. Some variation between budgeted and actual costs can be expected because most variable operating expenses vary with sales levels that cannot be predicted perfectly. The variances that occur can, however, tell operators a great deal about operational efficiencies, and experienced operators know that a key to ensuring profitability is to properly examine and manage controllable costs.

As shown in Figure 12.4, Randy's cost of food in dollars were lower than budgeted ($56,000 budgeted vs. $53,800 actual) as he would expect given his food sales shortfall, but his cost of sales percentage was over-budget (35.0% budgeted vs. 35.7% actual), an indication that his kitchen was not as cost-efficient as he would have liked. The same was true in his beverage sales category. As a result, a close examination of his product cost percentages will be an important part of Randy's analysis of budgeted versus actual expenses.

As was true in this example of Randy's Restaurant, in some cases, operators find that their product costs, when expressed as a percentage of sales, are too high, and they must be reduced. Operators facing this situation can choose from a variety of solutions to this problem if they first carefully consider the cost of sales equation.

Recall from Chapter 6 that the formula operators use to calculate their food cost percentage is:

$$\frac{\text{Cost of Food Sold}}{\text{Food Sales}} = \text{Food Cost \%}$$

Similarly, the formula used for calculating a beverage cost percentage is:

$$\frac{\text{Cost of Beverage Sold}}{\text{Beverage Sales}} = \text{Beverage Cost \%}$$

In its simplest terms, both of these product cost percentages can be expressed algebraically as:

$$\frac{A}{B} = C$$

where:

A = Cost of Products Sold
B = Sales
C = Cost Percentage

Figure 12.5 shows how algebra affects the equation and what the rules of algebra communicate to foodservice operators.

| Algebraic Rule | Foodservice Operator Takeaway |
| --- | --- |
| If A is unchanged and B increases, C decreases. | If product costs can be kept constant while sales increase, the product cost percentage goes down. |
| If A is unchanged and B decreases, C increases. | If product costs remain constant but sales decline, the cost percentage increases. |
| If A increases at the same proportional rate B increases, C remains unchanged. | If product costs go up at the same rate that sales go up, the cost percentage will remain unchanged. |
| If A decreases and B is unchanged, C decreases. | If product costs can be reduced while sales remain constant, the cost percentage goes down. |
| If A increases and B is unchanged, C increases. | If product costs increase with no increase in sales, the cost percentage will go up. |

**Figure 12.5** Rules of Algebra and Foodservice Operator Takeaways

In general, foodservice operators must control the variables that impact the product cost percentage and reduce the overall value of "C" in the product cost percentage equation. Basic product cost reduction approaches can be used to optimize their overall product cost percentage. These approaches are to:

✓ Ensure all purchased products are sold
✓ Decrease portion size relative to selling price
✓ Vary recipe composition
✓ Alter product quality
✓ Achieve a more favorable sales mix
✓ Increase selling price relative to portion size

These strategies can be applied when there are excessive food and/or beverage costs. It is the careful selection and mixing of these approaches to cost control that differentiates successful operators from their less successful counterparts.

It is not the authors' contention that there are no other possible product cost reduction methods. However, the alternative approaches presented illustrate how a careful analysis of cost percentage reduction alternatives can benefit all operators.

### ✓ Ensure All Purchased Products Are Sold

This strategy has tremendous implications, and it includes all phases of professional purchasing, receiving, storage, inventory, issuing, production, service, and cash control. Perhaps the hospitality industry's greatest challenge in product cost control is ensuring that all products, once purchased, generate cash sales that are ultimately deposited into an operation's bank account.

### ✓ Decrease Portion Size Relative to Selling Price

Product cost percentages are directly affected by an item's portion cost (see Chapter 6), which is a direct result of portion size. Too often, foodservice and bar operators assume their standard portion sizes must conform to some unwritten rule of uniformity. However, most guests would prefer a smaller portion size of higher-quality ingredients than the reverse. In fact, one problem some restaurants must address are that their portion sizes are too large. The result: excessive food loss because uneaten products left on plates must be thrown away. It is important to remember that foodservice operators solely determine portion sizes, and they are variable.

An example of the impact of portion size on the product cost percentage is shown in Figure 12.6. This figure presents the significant effect on the liquor cost percentage of varying the standard drink size assuming:

1) $21.00 per liter as the standard cost of the liquor
2) A 0.8-ounce evaporation rate per liter, resulting in 33 ounces of salable liquor per liter
3) A standard $10.00 selling price per drink

| Drink Size | Drinks Per Liter | Cost Per Liter | Cost Per Drink | Sales Per Liter | Liquor Cost % Per Liter |
|---|---|---|---|---|---|
| 2 oz. | 16.5 | $21.00 | $1.27 | $165.00 | 12.7% |
| 1¾ oz. | 18.9 | 21.00 | $1.11 | $189.00 | 11.1% |
| 1½ oz. | 22.0 | 21.00 | $0.95 | $220.00 | 9.5% |
| 1¼ oz. | 26.4 | 21.00 | $0.80 | $264.00 | 8.0% |
| 1 oz. | 33.0 | 21.00 | $0.64 | $330.00 | 6.4% |

**Figure 12.6** Impact of Drink Size on Liquor Cost Percentage at Constant Selling Price of $10.00 per Drink

Note in this example that the operator's liquor cost percentage ranges from 6.4% of sales when a 1-ounce portion is served to a 12.7% product cost when a 2-ounce portion is served.

Portion sizes of both food and drink items directly affect product cost percentages, and the guests' perceptions of value delivered. As a result, when establishing portion sizes, operators should carefully consider all variables affecting their sales. These may include location, service levels, competition, and the type of clientele being served.

✓ **Vary Recipe Composition**

Experienced foodservice operators know that even the simplest standardized recipes can often be changed. For example, consider the proper ratio of clams to potatoes when preparing 100 servings of high-quality clam chowder. Since the cost of one pound of clams far exceeds the cost of one pound of potatoes, 100 servings of clam chowder made with an increased amount of clams will cost more to produce than 100 servings of chowder made with increased amounts of potato. What constitutes an ideal recipe composition in this example? The answer must be addressed by the operator, and the "answer" in each standardized recipe used will impact the operation's overall food cost percentage.

Similarly, the proportion of alcohol to mixer affects beverage cost percentages. Sometimes the amount of alcohol in drinks can be reduced, and overall drink sizes can be increased. One example: Increasing the drink's proportion of lower cost standardized drink recipe ingredients such as milk, juices, and soda may enable the use of a smaller portion of higher cost spirit products. Utilization of this beverage cost reduction strategy often contributes to a feeling of satisfaction by the guest, while allowing the operator to reduce beverage cost percentages and increase profitability.

✓ **Alter Product Quality**

In nearly all cases, higher-quality food and beverage products cost more than lower-quality products. Therefore, one way to reduce product costs is to reduce product quality. This area must be approached with great caution because a foodservice operation should never serve products of unacceptable quality. Rather, cost conscious operators should always purchase the quality of product appropriate for its intended use.

When operators determine that an appropriate ingredient, rather than the highest-cost ingredient, provides good quality and value to guests, product costs might be reduced with product substitution. One caveat: Lower-quality products may cost less, but customers may perceive that menu items made from these lower-quality ingredients provide reduced levels of value.

✓ **Achieve a More Favorable Sales Mix**

Experienced operators know that their customers' menu item selection decisions have a direct and significant impact on product cost percentages. The reason: An operation's overall product cost percentage is determined in large part by the operation's **sales mix**.

Sales mix affects the overall product cost percentage anytime guests have a choice among several menu selections, and each selection has its own unique product cost percentage.

**Key Term**

**Sales mix:** The series of individual guest purchasing decisions that result in a specific overall food or beverage cost percentage.

To consider how sales mix directly affects an operation's overall product costs, assume that only three different menu items are sold in this operation. Each item is priced separately, but the operation also uses bundling (see Chapter 10) to produce a "Bundle Meal" that includes one of each item when purchased at the same time as shown in Figure 12.7.

When reviewing Figure 12.7, it is easy to see that if, on a specific day, 100% of the operation's guests bought a hamburger and nothing else, the operation's overall product cost would be 37.6%. If, on another day, 100% of its customers purchased a soft drink and nothing else, the product cost percentage for that day would be 15.2%.

| Menu Item | Product Cost | Selling Price | Product Cost % |
|---|---|---|---|
| Hamburger | $1.50 | $3.99 | 37.6% |
| French fries | $0.50 | $1.99 | 25.1% |
| Soft drink | $0.15 | $0.99 | 15.2% |
| Bundle Meal | $2.15 | $6.49 | 33.1% |

**Figure 12.7** Four-item Menu

Similarly, if every guest purchased only the "Bundle Meal" on a specific day, the operation would achieve a 33.1% product cost. An operation's actual product cost is largely determined by the "mix" of the individual product costs resulting from the menu item choices made by the operation's guests.

Operators can directly influence guest selection and sales mix by techniques including strategic pricing, effective menu design, and creative marketing. However, the guests will always determine an operation's overall product cost percentage because they produce an operation's unique sales mix.

Recognizing the impact of sales mix can help operators better understand that effective marketing and promotion of lower-cost items can reduce their product cost percentages and increase profitability, while allowing portion size, recipe composition, and product quality to remain constant.

✓ **Increase Selling Price Relative to Portion Size**

Some operators facing rising product costs and increased cost percentages think the only appropriate response is to increase their menus' selling prices. While this can be done, this tactic must be approached with great caution. There may be no bigger foodservice temptation than to raise prices to counteract management's ineffectiveness at controlling product costs!

There are times when selling prices must be increased, and this is especially true when there is inflation and when there are unique product shortages. Price increases should be considered, however, only when all other alternatives and needed steps to control product costs have been considered and effectively implemented. For example, assume an operation has a fresh orange slice as part of a garnish on a popular fresh salad. If the cost of oranges increases significantly, perhaps the garnish can be made with another fruit during the time oranges are very expensive.

Returning to the example of Randy's Restaurant, a review of Figure 12.4 reveals that Randy's total dollar amount spent for labor (management, staff, and employee benefits) was lower than budgeted ($70,000 budgeted vs. $66,960 actual), but his reduced sales level resulted in a slight increase in total labor percentage (35.0% budgeted vs. 35.2% actual).

Note also in Figure 12.4 that Randy's fixed costs (e.g., occupancy, depreciation, and interest) did not vary in dollar amount. However, because revenues did not reach their budgeted levels, these also represented a higher actual cost percentage of sales because the fixed dollars were spread over a smaller revenue base.

The marketing expense category in Figure 12.4 is an expense line item that illustrates the need to carefully compare budgeted expense to actual expense. Randy's September marketing budget was $3,212. His actual marketing expense was $1,350. While this variance might initially seem to be positive (less was spent on this expense than was previously budgeted), it is possible that some of Randy's shortfall in revenue may have come from the fact that he reduced his marketing expense. Experienced foodservice operators know that savings achieved in marketing costs often result in no savings at all!

Note also that Controllable Income in Figure 12.4 was budgeted at 21.4% of total revenue, yet the actual results were lower (21.2%). The difference in the dollar amount of Controllable Income achieved was $2,610 ($42,805 budgeted − $40,195 actual = $2,610 variance). Randy must decide if this constitutes a "significant" variance, and if so, it should be analyzed using the four-step "Operating Budget Monitoring Process" presented earlier in this chapter.

### Profit (Net Income) Analysis

A foodservice operation's actual level of profit is measured either in dollars, percentages, or both, and is the most critical number that most operators evaluate. Returning to Figure 12.4, it is easy to see that Randy's actual net income for the month was $14,696 or 7.7% of total sales. This is less than the $16,654 (8.3% of total sales) that was forecast in the operating budget. This reduced profit level can be tied directly to Randy's lower-than-expected sales.

When a business is unable to meet it topline revenue (see Chapter 1) forecasts, it typically means the budget was ineffectively developed, internal/external conditions have changed, and/or the operation's marketing efforts were not effective. Regardless of the cause, when sales do not reach forecasted levels, corrective action may be needed to prevent even more serious problems including significant future profit erosion.

Since foodservice profits (net income) are routinely reported as both dollars and percentages, foodservice operators may disagree about the best way to evaluate their profitability. One frequently hears comment in the hospitality industry such as "You bank dollars, not percentages!" To better understand this statement, note in Figure 12.4 that Randy's net income was a lower percentage than the operating budget forecast (7.7% actual net income vs. 8.3% budgeted net income).

However, consider the hypothetical results that would be obtained if Randy's actual net income percentage in September had been 8.5%. This would have been a *higher* percentage than his operating budget predicted (8.5% actual vs. 8.3% budgeted).

In this hypothetical scenario, the $16,150 total dollars of profit that would be generated ($190,000 actual sales × 8.5% net income = $16,150) would still be *less* than the initially budgeted 8.3% profit of $16,654. This, then, is the reasoning behind the statement, "You bank dollars, not percentages!"

## Modifying the Operating Budget

The best operators know their operating budgets should be an active and potentially evolving document. The budget should be regularly reviewed and, when necessary, modified as new and better information replaces that available when the original operating budget was developed. This is especially true when new data significantly (and perhaps permanently) affects the sales and expense assumptions used to create the budget. The operating budget should be reviewed anytime an operator believes the assumptions upon which it was based are no longer valid.

To illustrate, assume a foodservice operator employs 25 full-time employees. Each employee is covered under the operation's group health insurance policy. Last year, the operator agreed to pay 50% of each employee's insurance cost and, as a result, paid $300 per month for every full-time employee. The total cost of the insurance contribution each month was $7,500 (25 employees × $300 per employee = $7,500).

When this year's budget was developed, the operator assumed a 10% increase in health insurance premiums. If, later in the year, it is determined that premiums will be increased by 20%, employee benefit costs will be much greater than that projected in the original operating budget. This operator now faces several choices:

✓ Modify the budget.
✓ Reduce the amount contributed per employee to stay within the budget.
✓ Change (reduce) health insurance benefits/coverage to lower the premiums that will be charged to stay within the original costs allocated in the operating budget.

Regardless of this operator's decision, the operating budget, if affected, must be modified. There are situations in which an operating budget should be legitimately modified; however, an operating budget should never be modified simply to compensate for management inefficiencies.

To illustrate, assume, for example, that a budgeted "total labor" percentage of 25% is realistic, and achievable for a specific operation. The operation's managers, however, consistently achieve budgeted sales levels but just as consistently greatly exceed the total labor cost percentage targets established by the operating budget. As this occurs, resulting profits are less than projected. In this case, the labor cost portion of the operating budget should not be increased (nor should menu prices be increased!) simply to mask management's inefficiencies in controlling costs in this area. Instead, if the goal of a 25% total labor cost is indeed reasonable and achievable, then the operation's managers must correct the problem that yields a higher labor cost percentage.

Properly prepared operating budgets are designed to be achieved, and foodservice operators must do their best to ensure this occurs. There are cases, however, when operating budgets must be modified, or they will lose their ability to assist managers with decision-making responsibilities. The following situations are examples of those that, if unknown and not considered when the original budget was developed, may require operators to consider modifying their existing operating budgets:

✓ Additions or subtractions from offerings that materially affect revenue generation (e.g., reduced or increased operating hours)
✓ The opening of a direct competitor
✓ The closing of a direct competitor
✓ A significant and long-term or permanent increase or decrease in the price of major cost items
✓ Franchisor-mandated operating standard changes that directly affect (increase) costs
✓ Significant and unanticipated increases in fixed expenses such as mortgage payments (e.g., a loan repayment plan tied to a variable interest rate), insurance, or property taxes
✓ A manager or key employee change that significantly alters the skill level of the facility's operating team
✓ Natural disasters including floods, hurricanes, or severe weather that significantly affects forecasted revenue
✓ Changes in financial statement formats or expense assignment policies
✓ Changes in the investment return expectations of the operation's owners

Operators should have detailed knowledge of their operation, and then they can do a good job comparing their actual performance to their planned (budgeted) performance to identify significant variances. When they do, they can use the cost control strategies and procedures presented in this book to identify problem areas. They can also take the corrective actions needed to ensure that they and their operations can successfully meet all of their planned financial goals and objectives.

---

**Technology at Work**

The preparation and monitoring of operating budgets in the foodservice industry can be a time-consuming process.

Today, however, there are a variety of software programs available in both PC and Mac formats to help the owners of foodservice businesses develop and monitor their operating budgets.

To review the features of some of these helpful tools, enter "budgeting software for restaurants" in your favorite search engine and view the results.

---

**What Would You Do? 12.2**

"Well, it just makes sense to me that we start with how much profit we need to make and then estimate the sales and expenses required to make that profit level. Otherwise, why are we buying the food truck?" said Roland.

*(Continued)*

Roland was talking to Catriona, his partner in a new joint venture in which Roland and Catriona were sharing the cost of buying a food truck that would feature English-style Fish and Chips. They had agreed to share the truck's cost and operating profits on a 50-50 basis, and they were now preparing their first year's operating budget.

"I don't know if I agree with that," said Catriona. "It seems to me we need to estimate our realistic sales levels first and then our operating costs to see what profit we're likely to generate in the first year. Just because we want to make a certain amount of profit doesn't mean that we will!"

**Assume you were Roland. Why do you think it makes sense to focus on profits first when preparing an operating budget? Assume you were Catriona. Why do you think it is important to focus on estimated sales levels when preparing a first year's operating budget?**

## Key Terms

| | | |
|---|---|---|
| Historical data | Employee benefits | Unfavorable variance |
| Discretionary income | Variance | Significant variance |
| Mixed cost | Favorable variance | Sales mix |

---

**Operator's 10-Point Tactics for Success Checklist**

Evaluate your need for, and the current status of, each of the following operational tactics. For those tactics you think are important, but not yet in place, develop an action plan for its implementation including who will be responsible for the tactic's completion and the target date by which it should be completed.

| | | | | If Not Done | |
|---|---|---|---|---|---|
| Tactic | Don't Agree (Not Done) | Agree (Done) | Agree (Not Done) | Who Is Responsible? | Target Completion Date |
| 1) Operator understands the importance to profits of developing an accurate operating budget. | ___ | ___ | ___ | | |
| 2) Operator knows and can identify the unique purposes of each of the three different types of operating budgets. | ___ | ___ | ___ | | |

| Tactic | Don't Agree (Not Done) | Agree (Done) | Agree (Not Done) | If Not Done | |
|---|---|---|---|---|---|
| | | | | Who Is Responsible? | Target Completion Date |
| 3) Operator can state the advantages that will accrue when they develop and use well-prepared and detailed operating budgets. | ___ | ___ | ___ | | |
| 4) Operator recognizes the importance of assessing prior-period operating results when preparing an operating budget. | ___ | ___ | ___ | | |
| 5) Operator recognizes the importance of utilizing realistic assumptions about the next period's operations when preparing an operating budget. | ___ | ___ | ___ | | |
| 6) Operator recognizes the importance of understanding the organization's financial objectives when preparing an operating budget. | ___ | ___ | ___ | | |
| 7) Operator knows how to create realistic sales forecasts when preparing an operating budget. | ___ | ___ | ___ | | |
| 8) Operator knows how to create realistic fixed, variable, and mixed cost expense forecasts when preparing an operating budget. | ___ | ___ | ___ | | |
| 9) Operator has a system in place for identifying significant variances between actual operating results and planned results. | ___ | ___ | ___ | | |
| 10) Operator recognizes the importance of corrective action and/or budget modification when actual operating results vary significantly from planned results. | ___ | ___ | ___ | | |

# Glossary

**28-day accounting period:** An accounting period that is four weeks (28 days) in length instead of a calendar month that has between 28 and 31 days. There are 13 four-week periods instead of 12 monthly periods in one year when using this system.

**ABC inventory control system:** An inventory control system utilizing aspects of both physical and perpetual inventory systems.

**Acceptance hours:** The hours of the day in which an operation will accept food and beverage deliveries.

**Accounting period:** A period of time (e.g., an hour, day, week, or month) in which an operator wishes to analyze the revenue and expenses of a business.

**Accounting:** The system of recording and summarizing a foodservice operation's financial transactions and analyzing, verifying, and reporting the results.

**Accounts payable (AP):** Money owed by the foodservice operation to suppliers or others that has not yet been paid. Most often referred to as "AP."

**Ambience:** The character and atmosphere of a foodservice operation.

**As needed (purchase point):** A system of determining the purchase point by using sales forecasts and standardized recipes to determine how much of an item should be held in inventory. Also referred to as "just-in-time."

**As purchased ("AP"):** The weight, amount, or count of a product as delivered to a foodservice operation.

**Auditor:** An individual or entity responsible for reviewing and evaluating proper operating procedures in a business.

**Average (mean):** The value arrived at by adding the quantities in a series and dividing the sum of the quantities by the number of items in the series. Also commonly referred to as a "mean."

**Average sale per guest:** The mean amount of money spent per guest during a defined accounting period. Also commonly referred to as "check average."

**Back office accounting system:** The accounting software used to maintain a business's accounting records that are not contained in its POS system. This typically includes items such as payroll records, accounts payable and receivable, taxes due and payable, and net profit and loss summaries, as well as all balance sheet entries.

**Back-of-house staff:** The employees of a foodservice operation whose duties do not routinely put them in direct contact with guests.

**Balance sheet:** A report that documents the assets, liabilities, and net worth (owners' equity) of a foodservice business at a single point in time. Also commonly called the Statement of Financial Position.

**Bank reconciliation:** A process performed to ensure that a company's financial records are correct. This is done by comparing the company's internally recorded amounts with the amounts shown on its bank statement. Any differences must be justified. When there are no unexplained differences, accountants can affirm that the bank statement has been reconciled.

**Beginning inventory:** The monetary value of all products on hand at the start of an accounting period. Beginning inventory is determined by completing a physical inventory.

**Beverage cost percentage:** The portion of beverage sales that was spent on beverage expenses for consumers.

**Billing scam:** An effort to trick an operation into paying for products or services that it did not order, that have little or no value, or that were never delivered. Also known as an "invoice scam."

**Bin card:** A line on a spreadsheet (or a physical card) that details additions to and deletions from a product's inventory level.

**Bonding (employee):** Protection from loss resulting from a criminal event involving employee dishonesty and theft such as loss of money, securities, and other property.

**Bundling:** A pricing strategy that combines multiple menu items into a grouping that is then sold at a price lower than that of the bundled items purchased separately.

**Cash receipts journal:** An accounting record used to summarize transactions related to cash receipts generated by a foodservice operation.

**Cherry picker:** A foodservice operator who buys only those items from a supplier that are the lowest in price among the supplier's competition.

**Collusion:** Secret cooperation between two or more individuals that is intended to defraud an operation.

**Comp:** The foodservice industry term used to indicate food or beverage items served free of charge to guests or provided to them at prices lower than the regular menu price. "Comp" is short for "complimentary."

**Consumer rationality:** The tendency to make buying decisions based on the belief that the decisions are of personal benefit.

**Content management system (CMS):** Computer software used to load content into a digital menu display system.

**Contract price:** A price mutually agreed upon by supplier and operator. This price is the amount to be paid for a product or products over a prescribed period of time.

**Contribution margin (CM):** The dollar amount remaining after subtracting a menu item's product costs from its selling price.

**Controllable income:** The amount of revenue remaining after an operation's prime costs and other controllable expenses have been subtracted from its total sales.

**Controllable labor expenses:** Those labor expenses under the direct influence of management.

**Cost of sales:** The total cost of the products used to make the menu items sold by a foodservice operation.

**Count (product):** A number used to designate product size or quantity when purchasing food items.

**Credit memo:** An addendum to a supplier's delivery invoice that reconciles differences between a delivery invoice and the product(s) received.

**Delivery invoice:** A supplier's record of what is being delivered and at what cost.

**Depreciation:** The allocation of equipment costs and other depreciable assets based on the projected length of their useful life.

**Desired profit:** The profit level that an operator seeks to achieve with a predicted quantity of revenue.

**Digital menu board:** An integrated system that uses hardware and software to display an operation's menu on an electronic screen; also commonly referred to as a digital display menu or digital menu board.

**Dining bubble:** A controlled environment used to house foodservice guests while they are dining outdoors.

**Discretionary income:**
**Money a household or individual has available to invest, save, or spend after necessities are paid.**
Examples of necessities include the cost of housing, food, clothing, utilities, and transportation.

**Edible portion ("EP") cost:** The cost of a product after it has been cleaned, trimmed, cooked, and portioned. Also referred to as a "recipe ready cost."

**Embezzlement:** A crime in which a person or entity intentionally misappropriates assets that have been entrusted to them.

**Employee benefits:** Any form of rewards or compensation provided to employees in addition to their base salaries and wages. Benefits offered to employees may be legally mandated or voluntary.

**Empowerment (employee):** An operating philosophy that emphasizes the importance of allowing employees to make independent guest-related decisions and to act on them.

**Ending inventory:** The monetary value of all products on hand at the conclusion of an accounting period.

**Exempt employee:** Employees exempted from the provisions of the Fair Labor Standards Act. These workers typically are paid a salary above a certain level and work in an administrative or professional position. The U.S. Department of Labor (DOL) regularly publishes a duties test that can help employers determine who meets this exemption.

**Expenses:** The price paid to obtain the items required to operate a business. Also referred to as "costs."

**Extended price:** The price per purchase unit multiplied by the number of units delivered. This refers to the total per product price as listed on a delivery invoice.

**Fast casual (restaurant):** A sit-down foodservice operation with no wait staff or table service. Customers typically order off a menu board and seat themselves or take the purchased food elsewhere.

**Favorable variance:** Better-than-expected performance when actual results are compared to budgeted results.

**FIFO (first-in first-out) inventory system:** An inventory management system in which products already in storage are used before more recently delivered products.

**Fixed average:** The average amount of sales or volume over a specific series or time. For example: the first month of the year or the first week of the month.

**Fixed cost:** An expense that stays constant despite increases or decreases in sales volume.

**Fixed payroll:** The amount an operation pays for its salaried workers.

**Food cost percentage:** The portion of food sales that was spent on food expenses.

**Fraud:** Deceitful conduct to manipulate someone into giving up something of value by (a) presenting as true something known by the fraudulent party to be false or (b) by concealing a fact from someone that may have saved him or her from being cheated.

**Free-pour:** Pouring liquor from a bottle without measuring the poured amount.

**Freezer burn:** A form of deterioration in product quality resulting from poorly wrapped or excessively old items stored at freezing temperatures.

**Front-of-house staff:** The employees of a foodservice operation whose duties routinely put them in direct contact with guests.

**Full-service restaurant:** A foodservice operation at which servers deliver food and drink offered from a printed menu to guests seated at tables and/or booths.

**Furniture, fixtures, and equipment (FF&E):** Movable furniture, fixtures, or other equipment that have no permanent connection to a building's structure.

**General ledger:** The main or primary accounting record of a business.

**Guest count:** The number of individuals served in a defined accounting period.

**Historical data:** Information about an operation's past financial performance including revenue generated, expenses incurred, and profits (or losses) realized.

**Ideal expense:** An operator's view of the correct (appropriate) amount of expense necessary to generate a given quantity of sales.

**Income before income taxes:** The amount of money remaining after an operation's interest expense is subtracted from the amount of its restaurant operating income. Also, a business's profit before paying any income taxes due on the profits.

**Income statement:** Formally known as "The Statement of Income and Expense," a report summarizing a foodservice operation's profitability including details regarding revenue, expenses, and profit (or loss) incurred during a specific accounting period. Also commonly called the profit and loss (P&L) statement.

**Interest expense:** The cost of borrowing money.

**Issuing:** The process of transferring food and beverage products from storage to production areas for use in a foodservice operation.

**Jigger:** A small cup-like bar device used to measure predetermined quantities of alcoholic beverages. These items are usually marked in ounce and portions of an ounce. Examples: one ounce or 1.5 ounces.

**Job description:** A statement that outlines the specifics of a particular job or position within a foodservice operation. It provides details about the responsibilities and conditions of the job.

**Job specification:** A listing of the personal characteristics and skills needed to perform the tasks contained in a job description.

**Kilowatt hour (kwh):** A measure of electrical usage.

**Labor cost percentage:** A ratio of overall labor costs incurred relative to total revenue generated.

**Labor dollars per guest served:** The dollar amount of labor expense spent to serve each of an operation's guests.

**Law of demand:** The law of demand holds that the demand level (number of units sold) for a product or service declines as its price rises and increases as the price declines.

**LIFO (last-in first-out) inventory system:** An inventory management system in which the most recently delivered products are used before products already in storage.

**Maintenance log:** A document that contains detailed information about equipment and the actions (e.g., oiling, filter replacement, and repair) that has been performed on it.

**Menu engineering:** A system used to evaluate menu pricing and design by categorizing each menu item into one of four categories based on its profitability and popularity.

**Mixed cost:** A cost composed of a mixture of fixed and variable components. Costs are fixed for a set level of sales volume or activity and then become variable after this level is exceeded. Also referred to as a semi-fixed or semi-variable cost.

**Mobile wallet:** A virtual wallet that stores payment card information on a mobile device. Mobile wallets provide a convenient way for a user to make in-store payments at merchants listed with the mobile wallet service provider. Also known as a digital wallet.

**Moment of truth:** A point of interaction between a guest and a foodservice operation that enables the guest to form an impression about the business.

**Net income:** A number calculated as revenue minus operating expenses including depreciation, interest, and taxes, among others. It is useful to assess how much revenue exceeds the expenses of operating a business in a defined time period.

**Non-cash expense:** Expenses recorded on the income statement that do not involve an actual cash transaction. Examples include depreciation and amortization, which are expenses, and an income statement entry reduces operating income without a cash payment.

**Non-controllable expenses:** Costs that, in the short run, cannot be avoided or altered by management decisions. Examples include lease payments and depreciation.

**Non-controllable labor expenses:** Those labor expenses not typically under the direct influence of management.

**Non-profit sector (foodservice):** Foodservice in an organization where generating food and beverage profits is not the organization's primary purpose. Also referred to as the "non-commercial" sector.

**Occupancy costs:** Costs related to occupying a space including rent, real estate taxes, personal property taxes, and insurance on a building and its contents.

**Onboarding:** The ways in which new employees acquire the necessary knowledge, skills, and behaviors to become productive members of a foodservice operation's staff.

**Open check:** A guest check that initially authorizes product issues from the kitchen or bar but for which collection of the amount due has not been made (the account is not closed). Therefore, the amount due has not been added to the operation's sales total.

**Operating budget:** An estimate of the income and expenses of a business over a defined accounting period. Also referred to as a "financial plan."

**Other controllable expenses:** Expenses (costs) that a foodservice operator can influence with increases or decreases based on business decisions. Examples include marketing costs and utility costs.

**Over (cash bank):** A cashier's bank that contains more than the amount of cash, credit or debit card, or other authorized payment amount based on the number of menu items sold.

**Par level:** The amount of a stored item that should be held in inventory at all times.

**Payroll:** The term commonly used to indicate the amount spent for labor in a foodservice operation. Used for example in "Last month our total payroll was $28,000."

**Percentage variance:** The change in sales (or expense) expressed as a percentage, and which results from comparing two different operating periods.

**Perishable inventory:** Inventory items that possess relatively short shelf lives.

**Perpetual inventory system:** An inventory control system in which additions to and deletions from total inventory are recorded as they occur.

**Physical inventory:** An inventory control system tool in which an actual (physical) count and valuation of all product inventory on hand is taken at the close of an accounting period.

**Point-of-sale (POS) system:** An electronic system that records foodservice customer purchases and payments and other operational data.

**Popularity index:** The proportion of guests choosing a specific menu item from a list of alternative menu items. The formula for a popularity index is:

$$\frac{\text{Total Number of a Specific Menu Item Sold}}{\text{Total Number of All Menu Items Sold}} = \text{Popularity Index}$$

**Portion cost:** The product cost required to produce one serving of a menu item.

**Price (Noun):** A measure of the value given up (exchanged) by a buyer and a seller in a business transaction.

For example: "The price of the chicken sandwich combo meal is $9.95."

**Price (Verb):** To establish the value to be given up (exchanged) by a buyer and a seller in a business transaction.

For example "We need to price the chicken sandwich combo meal."

**Pricing unit:** The measure used to establish the selling price of a product. For example, pound, quart, case, bushel, or bag.

**Prime cost:** An operation's cost of sales plus its total labor costs.

**Product inventory:** All of the stored food and beverage items used to produce an operation's menu items.

**Product specification (spec):** A detailed description of a recipe ingredient or complete menu item to be served.

**Product yield:** The servable amount of a raw ingredient remaining after it has been cleaned, trimmed, cooked, and portioned.

**Productivity (worker):** The amount of work performed by an employee within a fixed time period.

**Profit:** The dollars that remain after all of a business's expenses have been paid. Also referred to as "net income."

**Profit margin:** The amount by which revenue in a foodservice operation exceeds its operating and other costs.

**Proprietary website:** A website in which the foodservice operator controls all the website's content and can readily make changes to it.

**Purchase order (PO):** A detailed listing of products a buyer is purchasing from a supplier.

**Purchase point:** The inventory level at which a stored item should be reordered.

**Purchase unit:** The packaging configuration (e.g., case, carton, bushel, box, or bag) by which a supplier would normally sell a product.

**Purchasing:** The process of "buying": placing an order, receiving a product (or service), and paying the supplier.

**QR code:** A QR (quick response) code is a machine-readable bar code that, when read by the proper smart device, allows inventory items to be identified by name, price, and delivery date.

**Quality:** Suitability for intended use. The closer an item comes to being suitable for its intended use, the more appropriate is its quality.

**Quick-service restaurant (QSR):** Foodservice operations that typically have limited menus and often include a counter at which customers can order and pick up their food. Most quick-service restaurants also have one or more drive-through lanes that allow customers to purchase menu items without leaving their vehicles.

**Ratio:** An expression of the relationship between two numbers; computed by dividing one number by the other number

**Receivable revenue:** The expected sales income when all food and beverage orders have been properly documented.

**Recipe conversion factor (RCF):** A mathematical formula that yields a number (factor) operators use to convert a standardized recipe that produces a known yield to the same recipe producing a desired yield.

**Recipe-ready (ingredient):** The form a recipe ingredient must be in before it is used in a standardized recipe.

**Refusal hours:** The hours of the day in which an operation will not accept food and beverage deliveries.

**Requisition:** The formal act of requesting a food or beverage product to be removed from storage for use in an operation.

**Restaurant row:** A street or region well-known for having multiple foodservice operations in a close distance from each other.

**Return on investment (ROI):** A measure of the ability of an investment to generate income

**Revenue:** The term indicating the dollars taken in by a business within a defined time period. Also referred to as "sales."

**Rolling average:** The average amount of sales or guest volume over a changing time period (for example, the last 10 days or the last three weeks). Also referred to as a "moving average."

**Safety stock:** Additions to working stock, held as a hedge against the possibility of extra demand for a given menu item. Note: This helps reduce the risk of being out of stock on any given inventory item.

**Salaried employee:** An employee who regularly receives a predetermined amount of compensation each pay period on a weekly or less frequent basis. The predetermined amount paid is not reduced because of variations in the quality or quantity of the employee's work.

**Sales forecast:** A prediction of the number of guests to be served and the revenues to be generated in a defined, future time period.

**Sales history:** A record of the number of guests served and the revenues generated in a defined time period.

**Sales mix:** The series of individual guest purchasing decisions that result in a specific overall food or beverage cost percentage.

**Sales per labor hour:** The dollar value of sales generated for each labor hour used.

**Sales to date:** The cumulative sales figures reported during a defined financial accounting period.

**Sales variance:** An increase or decrease from previously experienced or predicted sales levels.

**Sales volume:** The number of units sold.

**Scaling (recipe):** The process of adjusting the yield of a standardized recipe.

**Service recovery:** The actions taken to correct the results of a poor customer service experience (moment of truth) and to regain customer loyalty.

**Shelf life:** The amount of time a product held in inventory retains peak freshness and quality.

**Short (cash bank):** A cashier's bank that contains less than the amount of cash, credit or debit card, or other authorized payment amount based on the number of menu items sold.

**Short (delivery item):** When a supplier does not deliver the quantity of an item ordered for the appointed delivery date.

**Significant variance:** A variance that requires immediate management attention.

**Split-shift:** A working schedule comprising two or more separate periods of duty in the same day.

**Standard operating procedure (SOP):** The term used to describe how something should be done under normal business operating conditions.

**Standardized recipe cost sheet:** A record of the ingredient costs required to produce a specific standardized recipe.

**Standardized recipe:** The instructions needed to consistently prepare a specified quantity of food or drink at an expected quality level.

**Table turnover rate:** A measure of the amount of time a party occupies a dining table over a specific period of time.

For example, if an operation served 15 guests who occupied 5 tables between 7:00 p.m. to 11:00 p.m., the turnover rate would be calculated as:

15 guests/5 tables = 3 turns over a period of hours

**Target market:** The group of people with one or more shared characteristics that an operation has identified as most likely customers for its products and services.

**Third-party delivery:** The use of a smartphone or computer application that allows customers to browse restaurant menus, place orders, and have them delivered to the customer's location. In nearly all cases, the requested orders are delivered by independent contractors retained by the company operating the third-party delivery app.

**Ticket time:** The amount of time required to fill a guest's order or respond to a guest's request.

**Time and temperature control for safety foods (TCS foods):** Foods that must be kept at a particular temperature to minimize the growth of food poisoning bacteria or to stop the formation of harmful toxins.

**Tip distribution program:** A system of tip payment that allows an operation to distribute a customer's tip from an employee who actually received it to others who also provided service to the customer.

**Tip pooling:** A tip system that takes the tips given to individual employees in a group and shares them equally with all other members of the group. Used, for example, when bartender tips given to an individual bartender are shared equally with all bartenders working the same shift.

**Tip sharing:** A tip system that takes the tips given to one group of employees and provides a portion of them to another group of employees. Used, for example, when server tips are shared with those who bus the server's tables.

**Topline revenue:** Sales or revenue shown on the top of a business's income statements.

**Total labor:** The cost of the management, staff, and employee benefits expense required to operate a business.

**Total sales:** The sum of food sales and alcoholic beverage sales generated in a foodservice operation.

**Unfavorable variance:** Worse-than-expected performance when actual results are compared to budgeted results.

**Uniform System of Accounts for Restaurants (USAR):** A recommended and standardized (uniform) set of accounting procedures used to categorize and report restaurant revenue and expenses.

**Uniform system of accounts:** Accounting standards used to help provide accuracy, reliability, uniformity, consistency, and comparability in reporting financial information.

**User-generated content (UGC) site:** A website in which content including images, videos, text, and/or audio have been posted online by the site's visitors.

Examples of currently popular UGC sites include Instagram, X, Facebook, TikTok, Pinterest, and YouTube.

**Value:** The amount paid for a product or service compared to the buyer's view of what they receive in return for the purchase price.

**Value proposition:** A statement that clearly identifies the benefits an operation's products and services will deliver to its guests.

**Variable cost:** An expense that generally increases as sales volume increases and decreases as sales volume decreases.

**Variable payroll:** The amount an operation pays for those workers compensated based on the number of hours worked.

**Variance:** The difference between an operation's actual performance and its planned (budgeted) performance.

**Wagyu beef:** Beef from a Japanese breed of cattle that is highly prized for its marbling and flavor. In the Japanese language, "Wa" means Japanese, and "gyu" means cow.

**Walk (bill):** Guest theft that occurs when a guest consumes products that were ordered and leaves without paying the bill. Also referred to as a "skip."

**Waste percentage:** A ratio obtained by dividing a product's loss amount by its as purchased ("AP") amount.

**Weighted average:** An average that combines data on the number of guests served and how much each has spent during a specific accounting period.

**Weighted contribution margin:** The contribution margin provided by all menu items divided by the total number of items sold. Weighted contribution margin is calculated as:

$$\frac{\text{Total Contribution Margin of All Items Sold}}{\text{Total Number of Items Sold}} = \text{Weighted Contribution Margin}$$

**Working stock:** The quantity of goods from inventory reasonably expected to be used between deliveries.

**Yield percentage:** A ratio obtained by subtracting an ingredient's waste percentage from 1.00. It refers to the amount of product available for use by an operation after all preparation-related losses have been considered.

**Yield test:** A procedure used to determine actual edible portion ("EP") ingredient cost. It is used to help establish actual per usable unit costs for a product that will experience weight or volume loss in preparation.

# Index